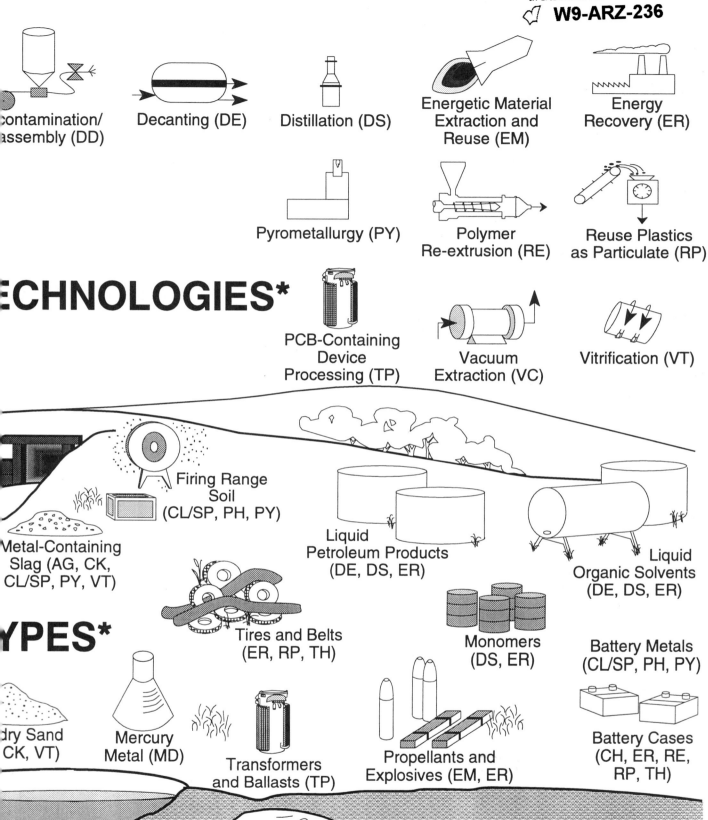

Decontamination/Disassembly (DD)

Decanting (DE)

Distillation (DS)

Energetic Material Extraction and Reuse (EM)

Energy Recovery (ER)

Pyrometallurgy (PY)

Polymer Re-extrusion (RE)

Reuse Plastics as Particulate (RP)

ECHNOLOGIES*

PCB-Containing Device Processing (TP)

Vacuum Extraction (VC)

Vitrification (VT)

Firing Range Soil (CL/SP, PH, PY)

Metal-Containing Slag (AG, CK, CL/SP, PY, VT)

Liquid Petroleum Products (DE, DS, ER)

Liquid Organic Solvents (DE, DS, ER)

Tires and Belts (ER, RP, TH)

Monomers (DS, ER)

Battery Metals (CL/SP, PH, PY)

YPES*

dry Sand CK, VT)

Mercury Metal (MD)

Transformers and Ballasts (TP)

Propellants and Explosives (EM, ER)

Battery Cases (CH, ER, RE, RP, TH)

Organics-Contaminated Soils and Sludges (DE, ER, SX, TD)

Vadose Zone VOCs (VC)

Fuel/NAPL (PR)

* The two-letter technology codes are shown below the applicable waste types.
** Includes a variety of aqueous processing technologies.

Recycling AND Reuse OF Industrial Wastes

LAWRENCE SMITH
Battelle

JEFFREY MEANS
Battelle

EDWIN BARTH
U.S. Environmental Protection Agency

BATTELLE PRESS
Columbus • Richland

The preparation of this document has been funded wholly or in part by the U.S. Environmental Protection Agency (EPA). This document has been reviewed in accordance with EPA's peer and administrative review policies and approved for publication. Mention of trade names or commercial products does not constitute endorsement or recommendation for use.

Library of Congress Cataloguing-in-Publication Data

Smith, Lawrence A.
    Recyling and reuse of industrial wastes / by Lawrence Smith, Jeffrey Means, Edwin Barth.
      p.   cm.
    Includes index.
    ISBN 0–935470–89–1 : $34.95
    1. Factory and trade waste—Recycling.   2. Hazardous waste site remediation.   I. Means, Jeffrey L.   II. Barth, Edwin F.   III. Title.
TD 897.845.S65   1995
628.4′458–dc20
                                    95-1591
                                    CIP

Printed in the United States of America

Additional copies may be ordered through:
Battelle Press
505 King Avenue
Columbus, Ohio 43201, USA
(614) 424–6393 or 1–800–451–3543

CONTENTS

Chapter 4 Product Quality Specifications

Chapter 5 Case Studies

FIGURES AND TABLES

FIGURES

TABLES

ABBREVIATIONS

AASHTO	American Association of State Highway and Transportation Officials	HWF	hazardous waste fuel
ABM	abrasive blasting media	LDPE	low-density polyethylene
AP	ammonium perchlorate	LIX	liquid ion exchange
API	American Petroleum Institute	LNAPL	light nonaqueous-phase liquid
ARRA	Asphalt Recycling and Reclaiming Association	MC	methylene chloride
ASTM	American Society for Testing and Materials	MEK	methyl ethyl ketone
		NAPL	nonaqueous-phase liquid
BF	blast furnace	NAS	Naval Air Station
BMC	bulk molding compound	NC	nitrocellulose
CERI	Center for Environmental Research Information	NCP	National Contingency Plan
		NEESA	Naval Energy and Environmental Support Activity
DBP	dibutyl phthalate	NG	nitroglycerine
DC	direct current	PAH	polycyclic aromatic hydrocarbon
DNAPL	dense nonaqueous-phase liquid	PC	polymer concrete
DNT	dinitrotoluene	PCB	polychlorinated biphenyl
ED	electrodialysis	PE	polyethylene
EDTA	ethylenediaminetetraacetic acid	PET	polyethylene terephthalate
EERC	Energy and Environmental Research Center	PM	polymer mortar
		PP	polypropylene
EPA	U.S. Environmental Protection Agency	PS	polystyrene
ERG	Eastern Research Group, Inc.	PVC	polyvinyl chloride
FHWA	Federal Highway Administration	RCRA	Resource Conservation and Recovery Act
HBX	high-blast explosive	RDX	research department explosive
HDPE	high-density polyethylene	REV	reverberatory furnace
HMX	high-melting explosive	RIM	reaction injection molding

RO	reverse osmosis	TCLP	Toxicity Characteristic Leaching Procedure
SMC	sheet molding compound	tetryl	2,4,6-tetranitro-N-methylaniline
SRK	short rotary kiln	TNT	trinitrotoluene
S/S	stabilization/solidification	TRI	Toxics Release Inventory
SSU	standard Saybolt unit(s)	TSCA	Toxic Substances Control Act
STLC	Soluble Threshold Limit Concentration	VOC	volatile organic compound
SVOC	semivolatile organic compound	WET	Waste Extraction Test

ACKNOWLEDGMENTS

This book was originally prepared for the Center for Environmental Research Information (CERI), Office Research and Development, of the U.S. Environmental Protection Agency. Edwin Barth: CERI, Cincinnati, Ohio; served as the Project Director and provided technical input and review. Funding was provided through Contract 68-C0-0068 with Eastern Research Group, Inc. (ERG), Lexington, Massachusetts; under the sponsorship of the U.S. Environmental Protection Agency. Paul Queneau of Hazen Research, Inc., Golden, Colorado; provided advice and consultation on methods of recycling metal-containing wastes. Line art was prepared by Loretta Bahn and Erin Sherer. Technical editing was provided by Lynn Copley-Graves. Heidi Schultz was the ERG project manger.

Preparation of this overview of waste recycling required information from a wide range of technical disciplines. The following Battelle staff members provided text for one or more chapters of the book.

Jody A. Jones
Arun Gavaskar
Karen Basinger
Prabahat Krishnaswami
Prakash T. Palepu
Manfred Luttinger
Bruce F. Monzyk
Mark A. Paisley

The following individuals provided a review of the document for CERI:

Ruth Bleyler
Hazardous Waste Division
U.S. Environmental Protection Agency–Region 1

John Blanchard
Office of Emergency and Remedial Response,
U.S. Environmental Protection Agency–
Headquarters

Albert Kupiec
AWD Technologies

Sally Mansur
Pollution Prevention Division
U.S. Environmental Protection Agency–Region 1

Raymond Regan
Associate Professor
Pennsylvania State University

Debbie Sievers
Waste Division
U.S. Environmental Protection Agency–Region 5

Mention of trade names or commercial products does not constitute endorsement or recommendation for use.

CHAPTER

1

INTRODUCTION

1.1 PURPOSE

The intent of this handbook is to assist pollution prevention efforts by encouraging recycling and reuse of wastes found on Superfund or Resource Conservation and Recovery Act (RCRA) Corrective Action sites. This handbook outlines specific technologies for recycling and reuse of materials that require remediation at contaminated sites. Case studies within the handbook document applications of these technologies to real-world conditions.

The main users of this handbook are expected to be personnel responsible for remediation of Superfund sites. Other potential users are personnel involved in RCRA corrective actions and environmental staff at facilities that generate industrial wastes.

This handbook is intended to increase the awareness of recycling options among personnel responsible for site cleanup by pointing out various technology options. The technologies in this handbook are described in generic terms; vendor-specific implementations of the technologies are not discussed due to the summary nature of the document. The document does not discuss the detailed costs or regulatory issues (except in the case studies in Section 5), as economics and regulations are complex, change rapidly, are location specific, and thus cannot be covered in a summary document. The economics and regulatory compliance of a technology option must be determined by site personnel based on local conditions. This document also does not cover risk evaluation to human health and the environment, which needs to be a consideration when remediating a site.

The general concept of recycling and reuse is straightforward—to find better ways to handle wastes other than depositing them in waste disposal sites. In practice, recycling takes on a variety of connotations, depending on the context and the user. For example, RCRA has specific regulatory definitions for recycling activities. Many publications make distinctions among terms such as *reclamation, recovery,* and *recycling* based on the processing required and the planned use. In this document, *recycling* and *reuse* are applied as general terms to indicate a range of activities, from direct reuse in a similar application to processing to produce a raw material for general use.

1.2 IMPETUS FOR RECYCLING AND REUSE

Recent Congressional legislation and U.S. Environmental Protection Agency (EPA) policy advocate pollution prevention, which includes environmentally sound recycling.

According to the National Contingency Plan (NCP) Preamble, Section 300.430(a)(1), EPA intends to focus available resources on selection of protective remedies that provide reliable, effective response over the long term. Recycling technologies already offer methods to remediate contaminants and minimize the amount of waste created. By

1

creating a small volume of residuals that require subsequent management, the cost effectiveness of a remedy may increase.

The NCP Preamble mandate for remedies that protect human health and the environment can be accomplished through a number of means, including recycling. The final rule indicates that alternatives shall be developed to protect human health and the environment by recycling waste or by eliminating, reducing, and controlling risks posed at each pathway at a site. The emphasis is clear: recycling is an approved means of site remediation.

Another criterion for remedy selection listed in the NCP is "reduction of toxicity, mobility, or volume through treatment." The regulation states that project managers should consider the degree to which alternatives employ recycling or treatment that reduces toxicity, mobility, or volume. Here again, emphasis is placed on consideration of remedies that use waste reduction to reduce the risks that a site poses.

1.3 SCOPE

This handbook outlines recycling and reuse approaches for a wide range of waste types. Both organic and inorganic contaminants in solid and liquid media are considered.

The following wastes containing mainly *organic* contaminants are discussed:

- Organic liquids
- Organic soils, sludges, and sediments
- Petroleum-contaminated soils, sludges, and sediments
- Solvent-contaminated soils, sludges, and sediments
- Propellants and explosives
- Rubber goods (e.g., tires and conveyor belts)
- Polymers
- Wire stripping fluff, plastic fluff, and paint debris

The following wastes containing mainly *inorganic* contaminants are discussed:

- Metal-containing solutions
- Metal-containing soils, sludges, and sediments

- Slags
- Mine tailings
- Ashes (bottom and fly)
- Spent abrasive blasting media
- Foundry sands
- Batteries
- Mercury-containing materials

In addition, the following *miscellaneous* wastes are covered:

- Chemical tanks and piping
- Structures
- Demolition debris
- Transformers and ballasts

This handbook does not cover wastes with established recycling markets because information on recycling these materials is available from other sources. Such wastes include:

- Municipal solid wastes, including nonleaded clear glass, white goods (e.g., refrigerators, washers, and dryers), automobiles, paper goods, and aluminum cans.
- Pure metals, including iron, steel, and ferrous alloys; copper and copper alloys; nickel and nickel alloys; and precious metals.
- Mixed metal wastes with over 40 percent metal content.
- Iron and steel blast furnace slags.

1.4 ORGANIZATION

The body of this handbook starts, in Chapter 2, with an illustration of recycling technology options and two summary tables to help the user quickly identify candidate recycling technologies for waste materials. The illustration presents the wide variety of waste types present at Superfund and RCRA Corrective Action sites, and shows how recycling and reuse options can be applied to these wastes. The illustration also provides a quick overview of the recycling potential of various waste materials. The first summary table lists wastes and shows possible recycling technologies for each waste. The second summary table outlines some of the key features of each technology.

The technologies shown in the second summary table are described in Chapter 3. Process description, schematic illustration, advantages, disadvantages or limitations, and operating features are summarized for each technology. These brief outlines familiarize the user with the technology. A listing of reference material for each technology provides sources of detailed information.

Chapter 4 reviews the general technical specifications (but not legal or regulatory requirements) of some typical end users of waste materials. The section describes the product characteristics and input material specifications for the more common users of wastes. The potential end-user of materials from a Superfund site will have a high level of concern about toxic contaminants and the associated potential for adverse health and safety effects or increased liability. There are few standards for recycled materials. Working with end-users to understand their process requirements and concerns is essential for developing workable specifications.

The application of recycling to real-world situations is examined through case studies. Chapter 5 describes several specific large-scale or commercial applications of recycling to waste materials. Site and waste type, technology application, recycling benefits, economics, and limitations are discussed for each case study.

2

COMPILATION OF TECHNOLOGIES AND APPLICATIONS

This section summarizes the wastes and technologies described in the following sections. The overall scope of the handbook is illustrated in Figure 2-1 (see inside front cover). The reader is encouraged to use Figure 2-1 as a starting point to identify technologies that are suitable for various waste materials. For situations that contain liquid wastes, such as lagoons, the reader should refer to the liquid waste types (shown as tankage) to identify the applicable technologies. The wastes and applicable recycling technologies are summarized in Table 2-1. The recycling technology characteristics are summarized in Table 2-2. The figure and summary tables will help users to quickly identify technology candidates applicable to wastes at their sites.

TABLE 2–1
Waste Types and Applicable Recycling Technologies

Waste Type	Applicable Recyling Technologies*
Wastes containing mainly organic contaminants	
Liquid organic solvent	• Distillation (3.1) • Energy recovery (3.2 and 3.3) • Decanting (3.4)
Liquid petroleum products	• Distillation (3.1) • Energy recovery (3.2 and 3.3) • Decanting (3.4)
Solvent-contaminated soils, sludges, and sediments	• Energy recovery (3.2 and 3.3) • Decanting (3.4) • Thermal desorption (3.5) • Solvent extraction (3.6)
Petroleum-contaminated soils, sludges, and sediments	• Energy recovery (3.2 and 3.3) • Decanting (3.4) • Thermal desorption (3.5) • Solvent extraction (3.6) • Use as asphalt aggregate (3.7)
Organic sludges	• Energy recovery (3.2 and 3.3) • Decanting (3.4) • Thermal desorption (3.5) • Solvent extraction (3.6)
Vadose zone volatile organic compounds (VOCs)	• In situ vacuum extraction (3.8)
Nonaqueous-phase liquids (NAPLs)	• Pump and recover (3.9)
Dissolved organics	• Freeze-crystallization (3.10)
Propellants and explosives	• Energy recovery (3.2 and 3.3) • Ingredient extraction, reuse, and conversion to basic chemicals (3.11–3.13)
Lead/acid battery cases	• Energy recovery (ebonite or polyethylene) (3.2 and 3.3) • Reuse as thermoplastic (polyethylene) (3.14–3.17)
Rubber goods (e.g., tires and conveyor belts)	• Energy recovery (3.2 and 3.3) • Size reduction and reuse (3.16) • Thermolysis (thermal conversion to basic hydrocarbon products) (3.17)
Liquid monomers	• Distillation (3.1) • Energy recovery (3.2 and 3.3)

(continued)

*Numbers in parentheses refer to the section(s) of this handbook where the technology is discussed.

TABLE 2–1 *(continued)*
Waste Types and Applicable Recycling Technologies

Waste Type	Applicable Recyling Technologies*
Wastes containing mainly organic contaminants *(cont.)*	
Solid polymers (low solids content)	• Energy recovery (3.2 and 3.3) • Reuse as construction material (3.7) • Re-extrusion (thermoplastics) (3.14) • Chemolysis (chemical conversion to monomers and oligomers) (3.15) • Size reduction and reuse (3.16) • Thermolysis (thermal conversion to basic hydrocarbon products) (3.17)
Solid polymers (high solids content, e.g., sheet molding compounds and bulk molding compounds [CaCO$_3$, glass fiber, and other inorganic filler in the 70% range])	• Reuse as construction material (3.7) • Size reduction and reuse (3.16) • Thermolysis (thermal conversion to basic hydrocarbon products) (3.17)
Paint residue and paint removal debris	• Energy recovery (3.2 and 3.3) • Thermolysis (thermal conversion to basic hydrocarbon products) (3.17)
Plastic fluff	• Energy recovery (3.2 and 3.3) • Thermolysis (thermal conversion to basic hydrocarbon products) (3.17)
Wastes containing mainly inorganic contaminants	
Low-concentration metals-containing solutions	• Chemical precipitation (3.18) • Ion exchange (3.19) • Liquid ion exchange (3.20) • Reverse osmosis (3.21) • Dialysis (3.22–3.23) • Evaporation (3.24) • Bioreduction (3.25)
High-concentration metals-containing solutions	• Freeze-crystallization (3.10) • Chemical precipitation (3.18) • Liquid ion exchange (3.20) • Evaporation (3.24) • Amalgamation (3.26) • Cementation (3.27) • Electrowinning (3.28)
Low-concentration metals-containing soils, sludges, and sediments	• Chemical leaching (3.29) • Vitrification (3.30)
High-concentration metals-containing soils, sludges, and sediments	• Chemical leaching (3.29) • Vitrification (3.30) • Pyrometallurgical processing • (3.31)
Silicate or oxide slags containing zinc, cadmium, lead	• Chemical leaching (3.29) • Pyrometallurgical processing or (oxide volatilization in a waelz kiln, flame reactor, or plasma furnace) (3.31)
Low-concentration metals-containing silicate/oxide slag, ash, dust, or fume	• Use as construction material (3.7) • Vitrification (3.30) • Cement raw materials (3.32)
High-concentration metals-containing silicate or oxide slag, ash, dust, or fume	• Chemical leaching (3.29) • Pyrometallurgical processing general) (3.31)
Abrasive blasting media	• Use as construction material (3.7) • Vitrification (3.30) • Cement raw materials (3.32) • Physical separation (3.33)
Foundry sand	• Use as construction material (3.7) • Vitrification (3.30) • Cement raw materials (3.32) • Physical separation (3.33)
Firing range soil	• Chemical leaching (3.29) • Pyrometallurgical processing (lead smelter) (3.31) • Physical separation (3.33)
Lead/acid battery internals	• Chemical leaching (3.29) • Pyrometallurgical processing (lead smelter) (3.31) • Physical separation (3.33)
Nickel/cadmium batteries	• Chemical leaching (3.29) • Pyrometallurgical processing 3.31)
Mercury-containing batteries	• Chemical leaching (3.29) • Roast and retort (3.34)
Mercury metal	• Mercury distillation (3.35)
Mercury-containing soils, sludges, and sediments	• Bioreduction (3.25) • Chemical leaching (3.29) • Physical separation (3.33) • Roast and retort (3.34)
Mercury-containing water	• Bioreduction (3.25)
Miscellaneous wastes	
Chemical tanks, pipes, and architectural materials	• Decontamination and disassembly (3.36) • Bulk metal reuse (3.36)
Nonmetal structures and demolition debris	• Use as construction material (3.7) • Decontamination and disassembly (3.36)
Wood debris	• Energy recovery (3.2)
Transformers and ballasts	• PCB flush and treat (dielectric) (3.37) • Metal recovery (electrical (3.37)

TABLE 2–2
Summary of Recycling Technology Characteristics*

Technology	Contaminant	Media	End Use	Limitations
ORGANIC LIQUID PROCESSING				
Distillation (3.1)	Organic solvents, petroleum, and monomers	Flowable liquids	Organic product	• Difference in boiling points • Impurities
Energy recovery (3.2 and 3.3)	Organic solvents, petroleum, monomers, and wood debris	Flowable liquids	Heating value in boiler, furnace, or cement kiln	• Energy content • Ash content • Impurities • May produce toxic byproducts • Waste moisture content • Explosive hazard
Decanting (3.4)	Organic solvents, petroleum	Two immiscible liquid phases	Organic product	• Typically produces a mixed product • Fluid density • Impurities
ORGANIC SOLIDS, SOIL, SLUDGE, AND SEDIMENT PROCESSING				
Energy recovery (3.2 and 3.3)	Solvents, petroleum, propellants and explosives, thermoplastic or thermosetting polymers, rubber goods, paint debris, or plastic fluff	Flowable soils, sludges, sediments, or particulates	Heating value in boiler, furnace, or cement kiln	• Energy content • Ash content • Impurities • May produce toxic byproducts • Waste moisture content • Explosive hazard
Decanting (3.4)	Solvent- or petroleum-contaminated soils, sludges, or sediments or organic sludges	Soils, sludges, or sediments	Organic liquid	• Typically produces a mixed product • Fluid density • Impurities
Thermal desorption (3.5)	Volatile or semivolatile organics	Soils, sludges, or sediments	Organic liquid	• Typically produces a mixed organic product
Solvent extraction (3.6)	Volatile or semivolatile organics	Soils, sludges, or sediments	Organic liquid	• Typically produces a mixed organic product
Soil vapor extraction (3.8)	Volatile organics	In situ vadose zone soils	Organic product	• Extraction/collection efficiency • Typically produces a mixed organic product
Pump and recover (3.9)	Nonaqueous-phase liquids (NAPLs)	In situ soils	Organic product	• Extraction/collection efficiency • Typically produces a mixed organic product
Propellant and explosive extraction (3.11)	Energetic materials	Munitions, rockets, etc.	Energetic materials	• Pretreatment step for reuse or chemical recovery • Difficult to extract safely
Propellant and explosive reuse (3.12)	Energetic materials	Munitions, rockets, etc.	Energetic materials	• Difficult to extract safely
Propellant and explosive conversion to basic chemicals (3.13)	Energetic materials	Munitions, rockets, etc.	Industrial chemicals (e.g., ammonium perchlorate, nitrates, phosphates)	• Difficult to extract safely

(continued)

*Numbers in parentheses refer to the section(s) of this handbook where the technology is discussed.

TABLE 2–2 *(continued)*
Summary of Recycling Technology Characteristics

Technology	Contaminant	Media	End Use	Limitations
ORGANIC SOLIDS, SOIL, SLUDGE, AND SEDIMENT PROCESSING *(continued)*				
Re-extrusion (3.14)	Thermoplastic	Solids with low concentrations of inert materials	Plastic products	• Most applicable to single polymer type • Color and opacity • Impurities
Chemolysis (3.15)	Polymers	Solids with low concentrations of inert materials	Organic chemicals,	• Most applicable to single polymer types
Size reduction and reuse (3.16)	Thermoplastic, thermosetting polymer, or rubber goods	Solids	Aggregate, bulk fillers, or filter media	• Type of polymer
Thermolysis (3.17)	Thermoplastic, thermosetting polymer, paint debris, plasticfluff, or rubber goods	Solids	Liquid or gaseous hydrocarbon feedstocks	• Inorganic filler or reinforcement content • Difficult to extract safely
WATER PROCESSING				
Freeze-crystallization (3.10)	Metals or dissolved organics	Water	Recovery of metals, metal salts, or organics	• Requires low concentrations of suspended solids
Precipitation (3.18)	Metals	Water	Recovery of metals or metal salts	• Typically requires additional processing to yield marketable product
Ion exchange (3.19)	Metals	Water	Recovery of metals or metal salts	• Typically requires additional processing to yield marketable product • Requires low concentrations of suspended solids and oil and grease
Liquid ion exchange (3.20)	Metals	Water	Recovery of metals, metal salts, or metal concentrates	• Requires low concentrations of suspended solids
Reverse osmosis (3.21)	Metals	Water	Recovery of metals or metal salts	• Typically requires additional processing to yield marketable product • Requires low concentrations of suspended solids and oil and grease
Diffusion dialysis (3.22)	Metals	Water	Recovery of metals or metal salts	• Requires low concentrations of suspended solids and oil and grease
Electrodialysis (3.23)	Metals	Water	Recovery of metals or metal salts	• Requires low concentrations of suspended solids and oil and grease
Evaporation (3.24)	Metals	Water	Recovery of metals or metal salts	• Energy-intensive process • Requires low concentrations of suspended solids and oil and grease
Bioreduction (3.25)	Mercury	Water	Recovery of mercury	• Mercury must be condenced and refined to produce a marketable product

(continued)

TABLE 2–2 *(continued)*
Summary of Recycling Technology Characteristics

Technology	Contaminant	Media	End Use	Limitations
WATER PROCESSING (continued)				
Amalgamation (3.26)	Mercury	Water	Recovery of mercury	• No net reduction in metal content • Mercury/metal amalgam must be retorted to obtain mercury metal
Cementation (3.27)	Metals	Water	Recovery of metals	• Requires low-cost source of less-noble metal
Electrowinning (3.28)	Metals	Water	Recovery of metals	• No net reduction in metal content
METAL-CONTAINING SOIL, SLUDGE, SEDIMENT, SLAG, OR OTHER SOLID PROCESSING				
Use as construction material (3.7)	Metals or inorganics	Petroleum-contaminated soils, slags, ashes, dusts, fumes, abrasive blasting media, foundry sand, or nonmetal demolition debris	Low-value structural product	• Leachable metals in waste
Bioreduction (3.25)	Mercury	Mercury-containing soils, sludges, or sediments	Recovery of mercury	• Typically requires additional processing to yield marketable product
Chemical leaching (3.29)	Metals	Soils, sludges, sediments, slags, ashes, dusts, fumes, firing range soils, batteries, or mercury-containing wastes	Recovery of metals or metal salts	• Leachable metals in treated residual • Volume of leaching solution required • Leaching solution must be regenerated and reused
Vitrification (3.30)	Metals or inorganics	Soils, sludges, sediments, slags, ashes, dusts, fumes, abrasive blasting media, or foundry sand	High- or low-value ceramic product	• Silica content • Leachable metals in product • Slagging conditions
Pyrometallurgical processing (3.31)	Metals, particularly cadmium, chromium, lead, nickel, and zinc at percent concentration	Soils, sludges, sediments, slags, ashes, dusts, fumes, firing range soils, or batteries	Recovery of metals	• Slagging • Water content • Arsenic • Halides
Feed to cement kiln (3.32)	Metals or inorganics	Slags, ashes, dusts, fumes, abrasive blasting media, or foundry sand	Cement	• Silica contenat • Iron content • Impurities
Physical separation (3.33)	Metals	Abrasive blasting media, foundry sand, firing range soils, lead/acid battery wastes, or mercury-containing soils, sludges, or sediments	Recovery of foundry sand or abrasive material; recovery of metals	• Typically requires additional processing to yield marketable product
Mercury roast and retort (3.34)	Mercury	Mercury-containing soils, sludges, sediments, or batteries	Recovery of mercury	• Halides • Water content
Mercury distillation (3.35)	Mercury	Free-flowing mercury liquid	Recovery of mercury	• Initial purity of waste mercury

(continued)

TABLE 2–2 *(continued)*
Summary of Recycling Technology Characteristics

Technology	Contaminant	Media	End Use	Limitations
MISCELLANEOUS WASTE PROCESSING				
Decontamination and disassembly (3.36)	Surface contamination	Chemical tanks, pipes, and architectural materials	Recovered bulk metals and construction materials	• Substrate value • Type and concentration of contaminant
PCB-containing transformer and ballast decontamination (3.37)	PCB-containing oil	Dielectric oil in electrical equipment	Recovery of oil and metals	• Thermal decontamination of metals can generate products of incomplete combustion

DESCRIPTION OF RECYLING TECHNOLOGIES

This section summarizes a wide range of recycling technologies that can be applied at Superfund or RCRA Corrective Action sites to obtain reusable materials from wastes containing organic and/or inorganic contaminants. The application of recycling technologies can increase the effectiveness of a remedial alternative by reducing the volume, toxicity, and/or mobility of hazardous substances, pollutants, and contaminants. Decreased site disposal costs and the value of the recycled product also may provide cost savings.

Each technology description includes seven sections:

- *Usefulness*, which summarizes the applicability of the particular technology for recycling.

- *Process Description*, which explains how the technology works.

- *Process Maturity,* which describes the technology's commercial availability and potential for implementation.

- *Description of Applicable Wastes,* which outlines the characteristics of waste streams typically processed for recycling using the technology.

- *Advantages,* which describes some of the particularly favorable aspects of the technology.

- *Disadvantages and Limitations,* which discusses potential challenges that the technology presents.

- *Operation,* which provides information on operating conditions and general implementation methods.

Wastes at Superfund or RCRA Corrective Action sites usually contain a mixture of contaminant and matrix types (mixtures of chlorinated and nonchlorinated organic sludges and soils, for example). These complex mixtures greatly increase the difficulty of processing to obtain a reusable product. At most sites, the application of several process options as a treatment train is required to encompass a remedial alternative for recovery of a useful product. Treatment trains often involve rough separation followed by separation and isolation. Rough separation is used to remove objectionable contaminants and to increase the concentration of the valuable constituents in the matrix. Separation and isolation further clean and upgrade the material to produce a useful product. Some common examples of rough separation followed by a separation and isolation process are:

- Thermal desorption or solvent extraction from soil, sludge, or sediment to produce a mixed organic liquid that is then purified and separated into reusable organic products by distillation.

- Precipitation to produce a filter cake followed by smelting, chemical leaching, and solution processing; or vitrification to produce useful materials.

The treatment train may require two separate facilities. For example, a small, onsite thermal desorption unit could be used to remove an organic conta-minant from soil or sludge. The recovered material could then be introduced into the physical separation and distillation operations of a commercial refinery along with the normal feedstock. The case studies described in Section 5 illustrate the application of several technologies used in sequence to form a treatment train.

3.1 DISTILLATION

3.1.1 Usefulness

Distillation is a thermal method that separates and concentrates volatile organic liquids from less volatile components to allow purification and reuse of one or more components. Desirable properties for feed material to petroleum distillation processes are given in Section 4.1. A case study of batch distilla-tion for onsite solvent recovery is described in Section 5.6.

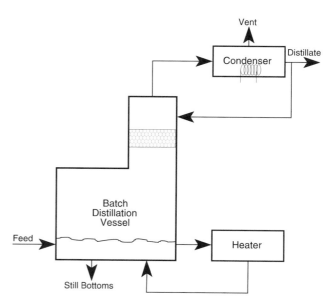

FIGURE 3-1. Batch distillation

3.1.2 Process Description

Distillation involves heating a liquid mixture of volatile compounds to selectively vaporize part of the mixture The vaporized material, which is en-riched with more-volatile compounds, is condensed and collected as distillate. The residual unvaporized liquid is enriched with less volatile materials. Distillation usually involves multiple stages of evap-oration and condensation to improve the separation of target compounds in the distillate and still bot-toms. Distillation may be done as a batch or as a continuous process. Small volumes of waste are best treated in batch stills, particularly if the composition varies or if the solids content is high (see Fig-ure 3-1). Continuous distillation processing can achieve higher throughput and can be more energy efficient but is more sensitive to the properties of the input materials (1).

3.1.3 Process Maturity

Distillation is a mature technology. The petro-leum and chemical industries have made extensive use of the process for many years. Small batch stills can conveniently recover small batches of spent sol-vents on site. Onsite solvent recovery units typically have a capacity in the range of 11.3 to 378 L (3 to 100 gal) per 8-hr shift (2).

3.1.4 Description of Applicable Wastes

Distillation is useful for recovering a wide variety of petroleum and organic solvents from liquid or-ganic wastes. For example, solvents can be recov-ered from wastes generated in paint formulation, metal cleaning and degreasing, or paint application (3).

Both the physical form and chemical content of the organic waste influence the ability to recover useful materials by distillation. Distillation is more effective if nonhalogenated and halogenated sol-vents are not combined in the wastes. Wastes with high solids content are not suitable for continuous distillation. Wastes containing organic peroxides or pyrophoric materials should not be processed by dis-tillation. Materials that polymerize can cause opera-tional problems.

3.1.5 Advantages

Distillation is a well-established process for re-covering useful materials from contaminated petro-leum and solvents. A solvent with a low boiling point (~100°C [212°F]) mixed with significantly less-volatile contaminants can be recovered with simple distillation equipment (4). Volatile residue loss in the still bottoms can be as low as a few per-cent. Distillation is technically able to reach any desired level of product purity, although practical and economic limits apply.

3.1.6 Disadvantages and Limitations

Distillation requires handling heated volatile organic liquids. Possible air emissions of volatile organics from process equipment and related storage tanks must be controlled.

Nonvolatile contaminants and low-volatility liquids remain in the still bottoms as viscous sludge. This sludge process residual must be managed, typically by incineration.

Distillation of complex mixtures of organics with similar boiling points requires expensive, complex equipment with high capital costs. However, purification of a volatile solvent contaminated with heavy oil and grease and of nonvolatile solids can be done in simple batch stills.

3.1.7 Operation

The main components of a distillation system are the heat source, a distillation vessel (batch) or column (continuous), and a condenser. The feed material is heated to vaporize volatiles, which are collected and condensed. Some of the condensed material usually is returned to the still to control distillate purity.

Distillation process equipment can cover an enormous range of size and complexity depending on the amount and type of material to be processed (3). Small quantities of contaminated solvents can be processed in simple batch stills. Large quantities of material usually are processed in continuous distillation columns. Dense, viscous, or high-solids materials require specialized equipment such as agitated thin film or wiped film heating systems (1, 5).

3.2 ENERGY RECOVERY (GENERAL)

3.2.1 Usefulness

A wide variety of organic wastes can be burned to recover energy in the form of steam or process heat. A description of desirable properties in feed materials for combustion to recover energy is given in Section 4.5.

3.2.2 Process Description

Energy recovery systems process waste containing organic materials in a boiler or other combustion device to recover energy values (see Figure 3-2). The organic component in the waste materials has the potential to serve as fuel in the combustion device and can, depending on the organic concentration, displace conventional fossil fuels such as oil or natural gas. Less concentrated organic materials require the use of a pilot fossil fuel to sustain combustion within the combustion device (6).

Energy value is recovered from energy recovery systems by generating steam or by using the hot flue gases produced for process heating. If temperatures within the combustion device are maintained above approximately 1,093°C (2,000°F), the more hazardous organic constituents also can be eliminated from the waste stream. Unlike incineration of the wastes, the major objective in an energy recovery system is the recovery of the steam or hot flue gas as a valuable product.

Inorganic portions of the waste materials exit the combustion device as ash that must be disposed of. If heavy metals are present in the ash, additional treatment may be necessary before disposal or reuse.

3.2.3 Process Maturity

Energy recovery systems are mature technologies and are available from many vendors. In some cases, existing combustion equipment can be used, with modification, to recover useful energy products from wastes containing organic materials. The scale of the equipment is limited only by the supply of material to be processed and the means of ash disposal.

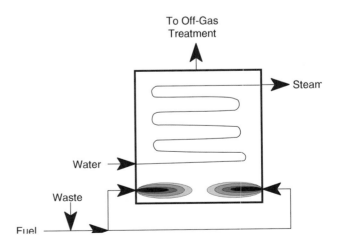

FIGURE 3-2. Energy recovery application

3.2.4 Description of Applicable Wastes

A wide variety of wastes can be processed and used for energy recovery. These include petroleum- or solvent-contaminated soils, propellants, rubber products, solid polymeric materials, automobile shredder residue, sludges, and wood debris (7). High-moisture materials such as sludges may limit the amount of energy that can be recovered from a particular waste, but any material with a measurable heating value over approximately 7,000 kJ/kg (3,000 Btu/lb) can be used for energy recovery. Halogenated solvents are poor candidates for energy recovery.

3.2.5 Advantages

Energy recovery allows for the generation of a useful product or products (steam, hot flue gas) from the waste materials. Depending on the design of the specific combustion device, little preparation of the feedstock is required, resulting in ease of operation.

3.2.6 Disadvantages and Limitations

Combustion processes may produce highly toxic products of incomplete combustion, such as dioxins and furans. The limitations of energy recovery as a general technology include the inability to process high-moisture wastes, such as sludges. In these cases, the attempted energy recovery is nothing more than incineration. Ash residue containing metals is another limiting factor in energy recovery systems. The ash must be treated as waste material, which in some cases means additional costs. If halogenated solvents are burned, corrosive acid vapors are introduced into the off-gas.

3.2.7 Operation

The specific operation of an energy recovery system varies with the type of combustion device. Boilers and similar systems usually are fueled at startup by natural gas or distillate oil; then, the waste material to be used as fuel is started and the startup fuel is turned off. The operation becomes routine and continues by feeding more waste to the boiler. Temperature is controlled by the air flowrate, fuel feed rate, and steam generation rate.

Other combustor types, such as fluidized beds, use a bed of inert material that receives the waste for combustion. These reactor systems are more flexible and can process a wider range of waste materials. Suspension burners require more tightly sized materials of generally small particle size, and are typified by pulverized coal combustion systems. These systems can require the addition of a pilot fossil fuel to stabilize the flame.

3.3 ENERGY RECOVERY (CEMENT KILNS)

3.3.1 Usefulness

Energy recovery can be particularly valuable when used in energy-intensive processes, such as the manufacture of Portland cement (see Figure 3-3). Due to the special characteristics for cement kiln combustion and the number of cement kilns permitted to burn hazardous wastes, energy recovery in cement kilns is discussed separately. A description of desirable properties for feed materials for combustion to recover energy appears in Section 4.5.

A large quantity of combustible waste is burned as fuel in cement kilns across the United States each year. The predominant waste fuels are hazardous solvents, waste oils, and tires. The number of cement kilns permitted to burn waste fuels has grown significantly since 1985. For example, in 1990, 6.8 percent of fuel consumption at cement kilns was hazardous waste fuel (9). EPA data suggest that 23.6 million metric tons (26 million tons) of hazardous waste fuel, with a heating value greater than 9,000 kJ (8,500 Btu), is available, but less than 10 percent of this fuel presently is committed to energy use (10, 11).

3.3.2 Process Description

Cement kiln operation is discussed in greater detail and is illustrated in Section 5.2. Raw materials such as limestone, clay, sand, and iron ore (perhaps supplemented by solid wastes of various types) are fed, either wet or dry, in specific proportions into the back (higher) end of a long rotary kiln. (Use of inorganic wastes as raw materials in cement kilns is discussed in Section 3.32.) Fuel is burned at the front (lower) end so that the hot combustion gas flow direction in the kiln is that of the solids. As the raw materials travel toward the front end of the kiln, they are heated, dehydrated, calcined, and then combusted and crystallized to form cement clinker. The process is extremely energy intensive, with maximum gas temperatures in excess of 2,200°C

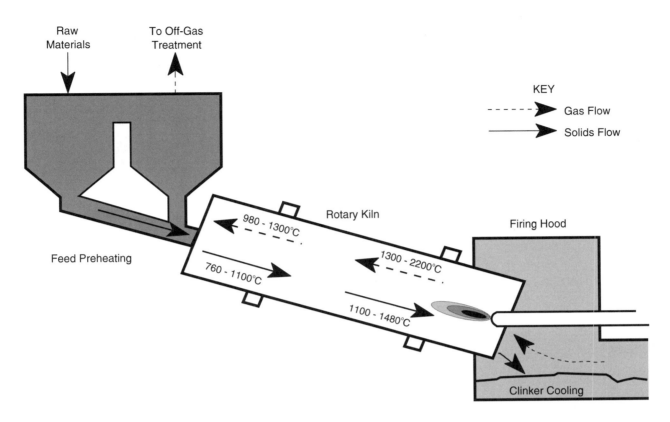

FIGURE 3-3. Energy recovery in a cement kiln (adapted from Gossman [8])

(3,990°F) at the front end of the kiln. This is a significantly higher temperature than in most hazardous waste incinerators, which typically operate at less than 1,480°C (2,700°F) and have shorter gas retention times than cement kilns. The extent of fuel combustion at cement kilns is greater than in most hazardous waste incinerators, with fewer emissions (8).

Depending on the waste fuel to be burned, pretreatment may be necessary, such as mixing, neutralization, drying, particle sizing, thermal separation or pyrolysis, and/or pelletization (8). Several different technologies are used to feed waste fuels into cement kilns, depending on the type of waste (9). Petroleum and petrochemical wastes generally can be pneumatically introduced.

3.3.3 Process Maturity

More than 25 cement kilns currently are permitted and actively burn hazardous waste fuels nationwide, with 10 additional plants soon to follow (9).

These 25 plants represent one-third of all cement plants in the country and one-quarter of clinker production capacity. At least seven cement kilns currently are burning tires or tire-derived fuel on an operating basis and another five on an experimental basis (12).

3.3.4 Description of Applicable Wastes

A wide variety of wastes can be recycled in this manner, depending on Btu content, physical characteristics, and chemical composition. Table 3-1 provides some guidance on the acceptability of different wastes based on physical characteristics and heat content. The Portland cement product is tolerant of a wide variety of trace constituents. As long as harmful constituents are controlled, destroyed, or rendered inert, the advantages of burning waste fuels are clear (8).

Typical constituents in hazardous waste fuel are xylene, toluene, mixed aliphatic hydrocarbons, acetone, methyl ethyl ketone, and a variety of chlori-

TABLE 3-1
Wastes Suitable for Treatment in a Cement Kiln (adapted from Grossman [8])

Friability		0% Organics ←──────────────────────────→ 100% Organics		
		<0.1% Organics	<5,000 Btu/lb	>5,000 Btu/lb
High friability ↑	Solids	Inorganic solids (see Section 3.32) Suitable for blending into raw feed	Organic-contaminated solids and sludges (such as contaminated soils or filter cake) Requires some form of thermal separation or direct feed for preheater	Grindable solid waste fuels (such as spent aluminum pot liner)
	Sludges	Inorganic liquids and sludges Suitable for blending into wet-process slurries; probably not suitable for dry-process kilns	Same as above for solids	Hazardous waste fuel (HWF) sludges Difficult to handle; can be blended into liquids or otherwise processed
↓ Low friability	Liquids	Same as above for sludges	Organic/water mixtures Suitable for incineration	Liquid hazardous waste fuels

nated solvents. Applicable solid wastes include tires, shredded plastic chips, petroleum industry residues, resins, and refuse-derived fuel (13). Cement kilns are very energy intensive. A single plant can potentially consume up to several million tires or several million kilograms of waste solvent and oil per year.

The most desirable waste fuel is relatively low in chlorine (Cl) content, is liquid, and has a moderately high Btu content, ranging from 25,600 to 41,900 kJ/kg (11,000 to 18,000 Btu/lb). The total suspended solids content of liquid fuels should be less than 30 percent to prevent plugging of the delivery system (14).

3.3.5 Advantages

Hazardous waste fuel generally burns cleaner than coal in a cement kiln and has lower associated nitrogen oxide (NO_x) and sulfur oxide (SO_x) emissions. The high temperatures achieved lead to thorough oxidation of the combustibles, and refractory contaminants such as nonvolatile metals are immobilized in the clinker's crystalline structure (8).

Kiln control is generally enhanced when burning even small quantities of hazardous waste fuel be-

cause the high level of volatiles stabilizes and aids combustion. The clinker acts as a scrubber for hydrochloric acid (HCl), and burning chlorinated solvents can enable the production of low-alkali cement, eliminating the need to purchase and add calcium chloride ($CaCl_2$) as an additional raw ingredient. In certain cases, burning hazardous waste fuel enhances cement clinker quenching and yields a product with higher strength and better grinding characteristics (8).

Steel from the reinforcing belts in tires does not need to be removed prior to burning because iron is an essential ingredient in Portland cement manufacture. Burning tires can actually reduce or eliminate the need to purchase iron ore to supplement the iron (Fe) content of quarry rock (12).

Financially, burning hazardous waste fuels at cement kilns can be profitable to both the waste generator and the kiln operator. The waste generator has reduced costs relative to other disposal options; the kiln operator is paid to burn fuel that would otherwise have to be purchased. In the case of tires, even if the cement plant operator pays up to 35 cents per tire, the economics still may be favorable to the operator (12).

3.3.6 Disadvantages and Limitations

Although a waste fuel recycling program dramatically reduces fuel costs for the plant operator, it also brings specific challenges:

- At less than 25,600 kJ/kg (11,000 Btu/lb), waste fuel is not a "hot" fuel; therefore, special attention must be paid to burner pipe design, optimum clinker cooler operation, and a tight hood seal (10, 11).

- Cement plant operators prefer to develop a uniform and consistent waste fuel supply so that processing parameters do not need constant adjustment.

- Solid and sludgy wastes present handling difficulties.

- Excessive Cl levels and the lack of compensating adjustments in kiln operations can lead to problems such as plugups, bad product, or kiln brick loss (8). Cl levels in the total fuel (compared with just the hazardous waste fuel) should be less than 3 percent by weight (14).

- Cement kilns must meet the stringent air standards specified in their permits. These standards affect the type of waste that can be burned.

- Cement kilns that burn hazardous waste fuel tend to produce a disproportionately large amount of cement kiln dust, which currently is exempted from the requirements of Resource Conservation and Recovery Act (RCRA) Subtitle C regulation under the Bevill Amendment, passed on October 12, 1980. Additional controls and possible regulation under Subtitle C, however, are being considered that could significantly affect the economics of burning hazardous waste fuels at cement kilns (9).

- Excessive levels of lead (Pb) and zinc (Zn) in the waste fuel can reduce product strength; excessive levels of Pb and chromium (Cr) can lead to safety hazards (8).

3.3.7 Operation

Cement kiln waste fuel recycling operation is quite simple for the generator. The waste fuel must be transported to the cement plant or to a permitted waste fuel processor (or "blender") as a broker for the kiln operator. Usually either the processor or the kiln operator performs pretreatment to prepare the waste fuel for burning.

3.4 DECANTING

3.4.1 Usefulness

Decanting is a physical method of separating two immiscible liquid phases to allow purification and reuse of one or more of the phases. A case study of decanting as part of a treatment train to recover petroleum from an oily sludge is described in Section 5.5.

3.4.2 Process Description

Decantation is used to remove small quantities of oil dispersed in water, or small quantities of water dispersed in oil (see Figure 3-4). Decantation relies on gravity to separate dense and light liquid phases. During the process, one liquid is dispersed as fine droplets in a second continuous phase. Decanting enhances the coalescence of the droplets of the dispersed phase into drops large enough to allow gravity to separate the two phases. Decanting efficiency increases with large droplets and large density differences between the phases and decreases with increasing viscosity of the continuous phase (15).

3.4.3 Process Maturity

Decanting to separate oil and water is a well-established technology. A variety of equipment types are available to efficiently treat a variety of oil and water mixtures.

3.4.4 Description of Applicable Wastes

Decanting is applicable to the separation of immiscible liquids. Oil can be recovered by treating contaminated oil, oily water, or oil sludges. Stable emulsions and suspensions must be broken to allow for efficient physical separation.

3.4.5 Advantages

Decanting allows for separation and recovery of oil from water or sludge. Parallel plate separation can reduce oil concentration in water from 1 percent to about 20 to 50 mg/L (1.2 to 2.9 grains/gal).

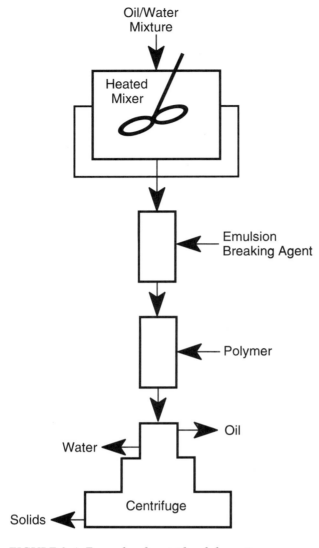

FIGURE 3-4. Example of centrifugal decanting

Dissolved air flotation provides about 90-percent effective removal of oil from water, with residual oil concentrations ranging from 90 to 200 mg/L (5.2 to 12 grains/gal) (15).

3.4.6 Disadvantages and Limitations

Surfactants or fine particulate will stabilize emulsions, greatly reducing the efficiency of decanting. Chemical additives are often needed to break stable emulsions (16). Decanting will only separate immiscible liquids such as oil and water. Further processing, such as distillation, is needed to separate mixtures of organics.

3.4.7 Operation

In its simplest form, a decanter is a tank with a large surface area to volume ratio. The continuous phase stands in the tank, while droplets of the dispersed phase combine and rise or sink (depending on density) to form a second phase that can be decanted. This simple approach is applicable only when the dispersed phase is present as large droplets and the speed and efficiency of separation is not critical.

Decanting works best when the surface area available for formation of a second phase is large compared with the volume of fluid. Use of corrugated parallel plates often yields a large surface area to volume ratio. A variety of commercial implementations of the parallel plate separator are available.

Coalescers, hydrocyclones, centrifuges, or air flotation units are used when the dispersed and continuous phases are difficult to separate or when a high proportion of solids are present. Coalescers provide a surface that enhances contact and agglomeration of the dispersed phase droplets, thereby improving phase separation. The surface can be a packed bed, a fiber mesh, or a membrane. The surface may be hydrophobic or hydrophilic, depending on the nature of the dispersed phase. Hydrocyclones and centrifuges improve phase separation by centrifugal action, induced by radial flow (hydrocyclones) or mechanical spinning (centrifuge). The mechanical centrifuge is particularly useful for separating light oil, water, and solids. Air flotation units improve phase separation by forming air bubbles or introducing them into the continuous phase. The bubbles provide a large surface area for collecting the dispersed phase droplets. The most common method used to form bubbles is to saturate water with air at elevated pressure and then to release the pressure (i.e., dissolved air flotation). Bubbles also can be introduced by gas sparging or electrolysis. Air flotation is used mainly when the dispersed phase is a low-density hydrophobic material, such as oil, and when the dispersed phase concentration is low.

3.5 THERMAL DESORPTION

3.5.1 Usefulness

Thermal desorption is a method used to physically recover volatile and semivolatile organic contaminants from soils, sediments, sludges, and filter cakes for reuse of the contaminant constituents. Volatile metals, particularly mercury, can be recovered by a thermal process similar to thermal desorption, called roasting and retorting (Section 3.34). A case study of thermal desorption as part of a treatment train to recover petroleum from an oily sludge is described in Section 5.5. A case study of thermal desorption to clean oily sand is described in Section 5.7.

3.5.2 Process Description

Thermal desorption systems heat the contaminated material to increase the rate of contaminant volatilization and cause the organic partition to the vapor phase (see Figure 3-5). The removal mechanisms are a combination of decomposition and volatilization. The organic-laden off-gas stream that volatilization creates is collected and processed. Unlike incineration, thermal desorption attempts to remove organics rather than oxidize them into their mineral constituents. As a result, thermal desorption systems operate at lower temperatures (95°C to 540°C [200°F to 1,000°F]).

3.5.3 Process Maturity

Thermal desorption processing equipment is in commercial operation and can be obtained readily from several vendors. Low-temperature treatment units are available as trailer-mounted or modular units, which can be transported to sites on standard highway transport trucks with a maximum gross vehicle weight of 36,300 kg (80,000 lbs). Thermal desorption has been selected for remediation of several Superfund sites (18).

3.5.4 Description of Applicable Wastes

Low-temperature systems have been used for the remediation of soil contaminated with a variety of volatile and semivolatile organic compounds (VOCs and SVOCs), including halogenated and nonhalogenated VOCs and SVOCs, polychlorinated biphenyls (PCBs), pesticides, and dioxins/furans (17, 19). The low-temperature desorption processes are best suited for removal of organics from sand, gravel, or rock fractions. The high-sorption capacity of clay or humus decreases the partitioning of organics to the vapor phase.

The heating process evaporates water as well as organics. Energy used to remove water from high-moisture-content wastes increases cost and does not assist in organic removal. Thermal desorption is therefore best applied to low-moisture-content wastes.

Thermal treatment units cannot process an unlimited range of particle sizes in the feed material. Units that use indirect heating require the presence of

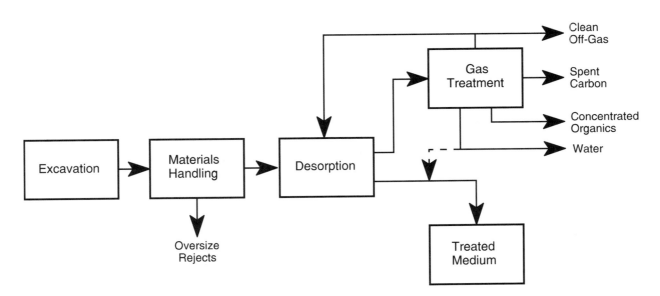

FIGURE 3-5. Example of thermal desorption process

smaller particles to provide sufficient contact surface with the heated wall. Fluidized bed or rotary kiln units require a reasonably narrow particle size range to control particle residence time in the heat zone. All units are unable to process large chunks due to heat transfer limitations and the potential for mechanical damage to the equipment from impact. The maximum allowed particle size depends on the unit but typically ranges from 3.8 to 5.1 cm (1.5 to 2 in.) in diameter.

3.5.5 Advantages

Thermal desorption allows for removal and recovery of organics from complex solid matrices. Desorption process conditions do not encourage chemical oxidation/reduction mechanisms, so combustion products are not produced. Thermal desorption treatment of low-organic-content streams is less energy intensive than incineration (20). In some cases, desorbed organics can be used directly; for example, desorbed petroleum hydrocarbons can be collected and used as a bitumen substitute in asphalt, or can be injected into a cement kiln or furnace for energy recovery. The operating temperature for thermal desorption reduces the partitioning of metals to the off-gas.

3.5.6 Disadvantages and Limitations

The successful performance of thermal desorption technology depends on the ability to maintain controlled heating of the contaminated matrix. The basis of the process is physical removal by volatilization. Organic removal is determined directly by the vapor pressure of the contaminant and the bed temperature. Treated waste retains traces of organic contaminants (20).

The organics stripped from the solid matrix are collected as a mixture, which must be distilled or otherwise purified before it can be reused as a solvent. Mixed petroleum products and nonhalogenated solvents can be used as fuel sources. Treated media typically contain less than 1 percent moisture. Dust can easily form during processing and when treated material is transferred out of the heating unit.

Thermal desorption is a capital-intensive operation that requires complex and expensive equipment. Costs can be controlled to some degree by matching processing equipment size to the amount of material to be treated. Low-temperature treatment requires complex equipment operating at elevated temperatures. Equipment operation involves hazards, but the nature and level of risk are consistent with industry practice.

3.5.7 Operation

Maximum temperatures and the heating systems used in commercial thermal desorption processing vary widely. Operating temperatures range from 95°C to 540°C (200°F to 1,000°F). Heating equipment includes rotary kilns, internally heated screw augers, externally heated chambers, and fluidized beds. Some systems use two-stage heating, where the first stage operates at low temperature to remove mainly water and the second stage operates at higher temperature to vaporize organics (21-23).

Most thermal desorption units use inert carrier gas to sweep volatilized organics away from the heated media. Treatment of off-gas from thermal desorption systems typically requires several steps. First, the hot off-gas is conditioned for efficient organic collection by removing particulate impurities. Various combinations of cyclone separators and baghouse filters remove the particulate impurities. Scrubbers and the processes of countercurrent washing and condensation then collect the organics. Most of the cleaned carrier gas is recycled to the heating unit, while carbon adsorption cleans the discharged portion.

3.6 SOLVENT EXTRACTION

3.6.1 Usefulness

The process of solvent extraction involves using an organic solvent to recover organic contaminants from soils, sludges, sediments, or liquids for reuse of the contaminant constituents.

3.6.2 Process Description

In solvent extraction, a solvent that preferentially removes the organic contaminant is contacted with the contaminated media (see Figure 3-6). Typical solvents include liquefied gas (propane or butane), triethylamine, or proprietary organic fluids. The extraction solvent is well mixed with the contaminated matrix to allow contaminants to transfer to the solvent. The clean matrix and solvent are then

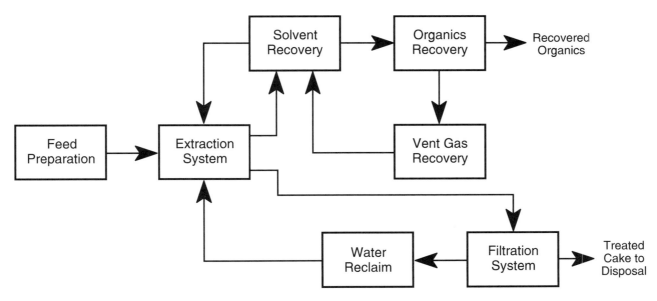

FIGURE 3-6. Example of the solvent extraction process (24).

separated by physical methods, such as gravity decanting or centrifuging. Distillation regenerates the solvent, which is then reused. Depending on their characteristics, recovered organic contaminants can be collected for reuse, processed to increase purity, or burned for energy recovery.

3.6.3 Process Maturity

Solvent extraction uses conventional solid/liquid contacting, physical separation, and solvent cleaning equipment. Two systems were tested as part of the Superfund Innovative Technology Evaluation (SITE) Demonstration Program (25, 26). Solvent extraction has been selected to remediate several Superfund sites and emergency response actions (18).

3.6.4 Description of Applicable Wastes

Solvent extraction is effective in treating sediments, sludges, and soils containing primarily organic contaminants such as PCBs, VOCs, halogenated solvents, and petroleum (27). Oil concentrations as high as 40 percent can be processed. Extraction is more effective with lower molecular-weight hydrophobic compounds. Contaminants targeted by solvent extraction include PCBs, VOCs, and pentachlorophenol (28).

3.6.5 Advantages

Solvent extraction recovers organic contaminants from an inorganic matrix, thus reducing the waste volume and preparing the organic for recycling. The treated residual is a dry solid. Solvent extraction can be used to treat wastes with high concentrations of organic contaminants.

3.6.6 Disadvantages and Limitations

Organically bound metals can transfer to the solvent along with the organics and restrict reuse options. Most extraction solvents are volatile, flammable liquids (20); as such, these liquid types require design and operating precautions to reduce risks of fire and explosion.

The liquid collected by solvent extraction processing of wastes typically contains a large number of different organics. A mixture composed of nonchlorinated organics may be suitable for energy recovery or asphalt-making. For higher-grade uses or when chlorinated organics also are present, further processing (e.g., distillation) may be required to separate the various organic liquids. Detergents and emulsifiers in the waste can reduce extraction performance. Water-soluble detergents dissolve and retain organic contaminants in the matrix. Detergents and emulsifiers promote foam formation, which complicates separation of the matrix and extraction solvent.

3.6.7 Operation

In solvent extraction processing, excavated waste materials are contacted with a selected extraction solvent. To be successful, the extraction solvent should have a high solubility in the contaminant and low solubility in the waste matrix.

As the process typically exhibits extraction behavior that is mass transfer limited, thorough mixing of the solvent and contaminated matrix is required. Some solvent extraction systems require the addition of water if the waste is a dry, nonflowing solid. In other systems, extraction fluid is added to make the waste flow.

The extraction solvent typically is purified by distillation. In systems that use pressurized solvents, such as liquefied gas or supercritical carbon dioxide (CO_2), vaporization occurs by pressure release, which causes the solvent to boil. With higher-boiling solvents, distillation tanks or towers may be used to separate the extraction solvent from the organic contaminants.

The triethylamine system extracts both water and organics. The contaminant/water/solvent mixture is heated to 55°C (130°F), where separate water and organic phases form. The phases are separated by decanting, and the contaminant and solvent are separated by distillation.

3.7 USE AS CONSTRUCTION MATERIAL

3.7.1 Usefulness

Low-value matrices with very low levels of leachable contaminants are suitable for reuse in construction applications after minimal processing. Sources for information on specifications for construction materials are given in Sections 4.11 and 4.12. A case study on the use of spent sand blasting media as aggregate in asphalt is described in Section 5.1.

3.7.2 Process Description

This category includes a collection of different processes that all use waste materials as an aggregate, usually in construction or road paving. Examples include the use of foundry sand, blasting sand, slag, fly ash, soil, or some other material as a blender aggregate in cement concrete, asphalt concrete (see Section 5.1), grading material, fill, or roadbed (see Figure 3-7). Alternatively, monolithic wastes such as plastic or elastomer wastes, bricks, other ceramics, mortars, or solidified wastes from stabilization/solidification (S/S) or vitrification projects can be crushed to form aggregate for the above purposes. These materials also can be reused in monolithic form for erosion control, diking material, artificial reefs, and other purposes.

Crushed stone also has agricultural applications (e.g., as a filler or conditioner in fertilizer and a mineral additive in animal feeds or poultry grit) and industrial applications (e.g., as an extender in plastic, rubber, paper, or paint). The principal requirements for the use of waste materials as aggregates or bulk materials are *acceptance*—by regulatory agencies, customers, and the public—and product *performance*. Typically the waste material must lend some useful function to the product and meet leach resistance criteria and specifications for physical properties (29). The "end use" should not simply be disposal in another form (termed "use constituting disposal" or "sham recycling"). Even if regulatory requirements and technical specifications are met, customers or the public may be reluctant to accept the use of those materials.

3.7.3 Process Maturity

The technology for this group of processes is mature and commercially available. A wide variety of materials have been used as aggregates in construction projects for many years.

FIGURE 3-7. Construction material loader

3.7.4 Description of Applicable Wastes

Applicable wastes include a wide variety of inorganic waste materials. Pavements, construction materials, ceramics, or glasses that are either aggregates or can be crushed to form aggregates are typical (30). Some fly ash and slag wastes can be used to supplement or replace Portland cement. Reuse usually takes place in the public domain, so wastes should contain low levels of relatively low-hazard contaminants.

3.7.5 Advantages

The structural properties of recycling aggregates make them well suited for the designed end uses. In addition, turning waste materials into aggregates conserves landfill space for higher-hazard waste materials and avoids disposal costs.

3.7.6 Disadvantages and Limitations

The main disadvantage of recycling aggregates is the risk or perceived risk of exposure to hazardous materials, which creates health concerns in the public. The two principal exposure pathways are inhalation of dusts or exposure to ground or surface water containing soluble metals that have leached from the aggregate. Any such recycling project should be able to demonstrate that no significant risk is added to either process or product. There should be negligible incremental risk to the recycling process workforce or to the public potentially exposed to the recycled material. Potential liabilities relating to the real or perceived health effects of the recycled material may exist for the waste generator.

Other limitations pertain to product specifications, such as strength, grading, chemical composition and purity, and chemical reactivity (31). Section 4.12 summarizes American Society for Testing and Materials (ASTM) specifications for aggregates and bulk construction materials; Section 5.1 describes a number of product acceptance criteria for recycling waste aggregates into asphalt concrete. Aggregate for landfill cover should have low dispersability; otherwise dusting will occur. Waste aggregate used to produce mortar or other cementitious products should have a low metallic aluminum content because aluminum corrodes and releases hydrogen gas (H_2), which decreases the strength of the cement.

3.7.7 Operation

This recycling technology is straightforward and involves little in the way of operation. Unless the reuse location is on site, the waste aggregate must be transported to the recycler's location. If the aggregate is going to be used as a construction material or as aggregate in concrete, crushing the waste aggregate and/or grading it by particle size may be necessary. Storage requirements in compliance with any pertinent regulations may involve an impervious liner, bins, or hoppers to prevent leaching. Special handling and worker protection may be required to minimize exposure to dust.

3.8 IN SITU VACUUM EXTRACTION

3.8.1 Usefulness

In situ vacuum extraction removes volatile organics from the vadose zone without bulk excavation. The extracted organics can be collected for reuse by condensation or adsorption/regeneration.

3.8.2 Process Description

In situ soil vapor extraction is the process of removing VOCs from the unsaturated zone (see Figure 3-8). Blowers attached to extraction wells alone or in combination with air injection wells induce airflow through the soil matrix. The airflow strips the VOCs from the soil and carries them to extraction wells. The process is driven by partitioning of volatile materials from solid, dissolved, or nonaqueous liquid phases into the clean air that the blowers introduce (32). Air emissions from the systems typically are controlled aboveground by adsorption of the volatiles onto activated carbon, by thermal destruction (incineration or catalytic oxidation), or by condensation through refrigeration (33).

3.8.3 Process Maturity

Vacuum extraction to remove volatile organics from the vadose zone is a mature and widely applied technology. A reference handbook on soil vapor extraction is available (34). Extracted organics can be recovered as useable liquid either by chilling to directly condense liquids or by adsorption onto (and subsequent regeneration of) carbon or other media. Recovery of liquids is technically feasible,

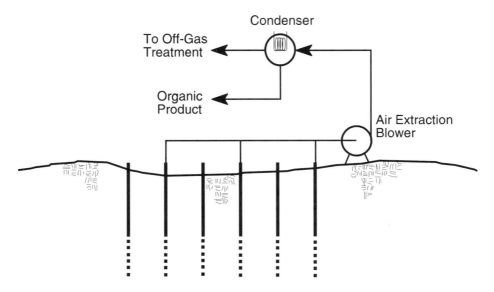

FIGURE 3-8. *Example of a vacuum extraction system.*

but treatment using thermal destruction is more frequently used.

3.8.4 Description of Applicable Wastes

Vacuum extraction has demonstrated its ability to remove halogenated and nonhalogenated VOCs and nonhalogenated SVOCs. The process also is potentially effective for halogenated SVOCs (35).

3.8.5 Advantages

Vacuum extraction allows recovery of organics without bulk soil excavation.

3.8.6 Disadvantages and Limitations

The organics collected by vacuum extraction typically contain a large number of different organics. In addition, the composition changes as extraction proceeds due to differences in the relative volatility of the organic contaminants. A mixture of nonchlorinated organics may be suitable for energy recovery or for making asphalt. For higher-grade uses or when chlorinated organics also are present, further processing (e.g., distillation) may be required to separate the various organic liquids.

Moisture in the extracted vapor stream freezes during condensation of organic vapors. High moisture content causes sufficient icing on condenser surfaces to significantly reduce the efficiency of chilling for collecting organic liquids.

3.8.7 Operations

Application of vacuum extraction relies on the ability to deliver, control the flow of, and collect stripping air. The main factors favoring application of vacuum extraction are the contaminant vapor pressure, the air conductivity of the soil, the soil moisture content, the sorption capacity of the soil, and the solubility of the contaminant in water. High vapor pressure, high conductivity, low soil moisture, low sorption, and low water solubility improve extraction efficiency (35).

3.9 PUMPING AND RECOVERY

3.9.1 Usefulness

The pumping and recovery process can be used to extract immiscible organic liquids in the subsurface, which can then be reused. A case study of pumping and recovery of coal tar wastes is described in Section 5.8.

3.9.2 Process Description

Many organic liquids have low solubility in water and can be present in natural formations as accumulations of nonaqueous-phase liquids (NAPLs). The location and configuration of the accumulations depend upon the density, interfacial tension, and viscosity of the NAPL. For NAPLs with a density lower than water (LNAPLs), the NAPL often is found as a layer floating on the top of the ground water. For NAPLs with a density higher than water (DNAPLs)—for example, some chlorinated solvents, organic wood preservatives, coal tars, pesticides, or PCBs—an organic liquid phase can pool on low-permeability geologic formations.

The accumulations of organic liquid can be recovered for reuse by installation of wells and pumps. Recovery of LNAPLs usually involves a skimming system that preferentially removes the floating organic liquid (see Figure 3-9). Ground-water pumping in the LNAPL recovery well can be used to depress the ground-water level, thus creating a gravity gradient to assist in transport of LNAPL to the skimming system. DNAPLs can be recovered by pumping from liquid deposits.

3.9.3 Process Maturity

Pump and recover systems use simple, commercially available well installation and pumping equipment and techniques (36). Pumping and collection of NAPLs are used to recover either light or dense organics at Superfund or RCRA Corrective Action sites (37, 38).

3.9.4 Description of Applicable Wastes

Pump and recover system can be applied to collect immiscible organic liquid deposits from in situ formations.

3.9.5 Advantages

Deposits of dense or light NAPL floating on ground water can contribute to ground-water and surface-water contamination. Pumping and recovery of NAPL pools represent a low-cost technology to collect and return organic liquids for reuse.

3.9.6 Disadvantages and Limitations

Placement, installation, and operation of wells must be done carefully to reduce the risk of promoting NAPL migration due to pumping operations.

The recovered NAPL typically is a mixture of several organics and water. Water and NAPL can be separated by decanting. In many applications, preprocessing is needed to break emulsions. A mixture of nonchlorinated organics may be suitable for energy recovery or making asphalt. For higher-grade uses or when chlorinated organics are present, fur-

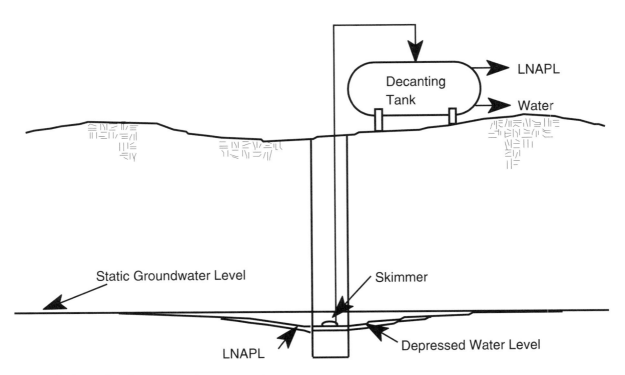

FIGURE 3-9. Example of a pump and recover system

ther processing (e.g., distillation) may be required to separate the various organic liquids.

3.9.7 Operations

Mobile NAPLs can be pumped from wells and drains. Systems may use one pump to withdraw only the NAPL or the NAPL mixed with water, or they may use two pumps, one to withdraw NAPL and another to withdraw water. Wells should be placed in stratigraphic traps to optimize recovery where NAPL pools are present. Long-term recovery is increased if maximum thickness and saturation of NAPL is maintained at well locations (36).

In typical pumping operations, the pumping rate is held constant. Cyclic pumping may be useful in situations where slow mass transfer rates reduce the availability of the contaminant. In cyclic pumping, the removal rate varies with time by alternating periods of pumping and no pumping. When pumps are idle, contaminants can flow out of more restricted low-permeability areas. The petroleum industry uses cyclic pumping to enhance oil extraction (32).

3.10 FREEZE-CRYSTALLIZATION

3.10.1 Usefulness

Freeze-crystallization is a physical method to recover concentrated solutions of organic or metal salt contaminants for reuse by processing high-concentration water solutions.

3.10.2 Process Description

Freeze-crystallization is a separation technique used to separate solids from liquids or liquids from liquids (see Figure 3-10). For hazardous waste treatment, the process is used to separate water from the hazardous components. Freeze-crystallization uses a

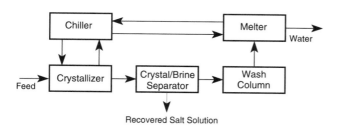

FIGURE 3-10. Example of the freeze-crystallization process (adapted from Heist [39]).

refrigeration process that causes water in a solution to form into crystals. The crystals are then separated from the remaining material, washed, and melted into a purified stream. This leaves a concentrated volume of the water-stripped material to be processed for resource recovery or, in some cases, used directly. Ferric chloride solution concentrated by freeze-crystallization treatment of acid pickling baths can be used in water treatment, for example.

3.10.3 Process Maturity

Development of the freeze-crystallization process began in the 1950s, when it was commercialized for fractioning p-xylene from its isomers. The process has been used to desalt seawater, to purify organic chemicals, and to concentrate fruit juices, beer, coffee, and vinegar. In the late 1980s, freeze-crystallization was applied to the treatment of hazardous wastewaters in the United States. The newer, direct-contact refrigeration cycles, in which coolant is mixed directly with the input solution, have improved the efficiency of the technology (40).

3.10.4 Description of Applicable Wastes

Freeze-crystallization is most effective when used to treat nonfoamy, nonviscous wastes and liquid wastes with low suspended solids content. Demonstrated applications include recycling of acid pickle liquor, recycling of alkaline baths used in metal finishing, and recovery of materials from ammunition plant wastewater (5).

3.10.5 Advantages

The freeze-crystallization process is energy efficient, closed (no emissions), and flexible enough to adjust to a wide variety of wastes. Input solutions generally do not require pretreatment. Contaminants in the crystal melt contain from 0.01 to 0.1 percent of the contaminants in the input solution. The residual retains the volatiles that are in the waste. Low temperatures during operations help avoid corrosion of metal equipment.

3.10.6 Disadvantages and Limitations

High capital costs are associated with the process, depending on the operation, facility size, and construction materials. The capital costs can be two or three times higher than the costs of evaporation or distillation systems. Production hangups can occur because of the complexity of the process used to control crystal size and stability. Eutectic conditions can occur where more than one material crystallizes at the same time. Buildup on the vessel walls of the crystallizer, or fouling, eventually occurs and requires system shutdown to allow foulant removal (41).

3.10.7 Operation

The major components of the process are crystallizers, separators, melters, and a refrigeration system. A solution is fed into one or more crystallizers, where the solution is either cooled or evaporated. Crystals begin to form and can be separated from the residual concentrate by filtration, hydrocyclones, centrifuges, or wash columns. The crystals are washed to remove additional concentrate, then melted. The residual is a concentrated contaminant stream that can be processed further for organic or metal recovery or, in some cases, can be reused directly. Processing capacities range from about 3.78 L/min (1 gal/min) in mobile units to 378 L/min (100 gal/min) in larger systems. The operating costs primarily reflect electricity and staffing needs.

3.11 PROPELLANT AND EXPLOSIVE EXTRACTION

3.11.1 Usefulness

Propellant and explosive extraction applies physical and chemical methods to remove energetic materials from metal casings for reuse, conversion to basic chemicals, or burning for energy recovery.

3.11.2 Process Description

Different energetic materials provide propulsive or explosive functions in rocket motors, munitions, and similar devices. These materials are made of different chemicals and have different characteristics of solubility, sensitivity to ignition, burn rate, and energy content. The potential for reuse varies widely depending on the physical form, chemical content, and reactive characteristics of the materials.

The energetic materials to be recovered may be present in obsolete devices or in contaminated soils, sludges, or manufacturing residues. Obsolete devices may be refurbished and reused for their original purpose, or may be disassembled so that the energetic materials can be removed. The removed energetic materials may be purified and reused (see Section 3.12), processed to recover useful chemicals (see Section 3.13), or burned for energy recovery (see Section 3.2.).

Munitions can contain various primers, igniters, propellants, explosives, and chemicals (see Figure 3-11). Used to initiate propellant burning, priming and igniting compounds often are high-sensitivity materials, such as metal azides or fulminates. Propellants (usually containing nitrocellulose, nitroglycerine, and/or nitroguanidine) burn rapidly to drive the projectile. A fuse and igniters trigger the projectile. The projectile may be filled with explosives, pyrotechnic mixtures, or smoke-generating chemicals. Explosive ingredients include trinitrotoluene (TNT), high-melting explosives, and ammonium nitrate. Chemicals frequently used in pyrotechnic and smoke mixtures include magnesium, zinc, and metal nitrates. The explosive or chemical fill is usually held in a binder. Bombs contain materials similar to projectiles but do not require propellants.

Rocket motors are thin metal casings containing an energetic material held in place with a binder. The energetic material typically is a mixture of ammonium perchlorate oxidizer and aluminum metal fuel, held by a polymer binder.

3.11.3 Process Maturity

Commercially proven methods are available to extract many types of energetic materials. Additional extraction methods are under development to improve on the efficiency of the existing methods and to allow extraction of energetic materials from previously unprocessable devices.

3.11.4 Description of Applicable Wastes

Recovery and reuse methods should be applied only to munitions and rocket motors that have documented histories. Documentation should include the method of manufacture and the composition of all energetic materials in the device. Propellants that contain combustion modifiers, such as lead compounds, are difficult to reuse because of the stringent controls on lead emissions. Primary explosives and initiating explosives, such as lead azide or metal fulminates, generally are not candidates for recovery and reuse due to their sensitivity. Pyrotechnic

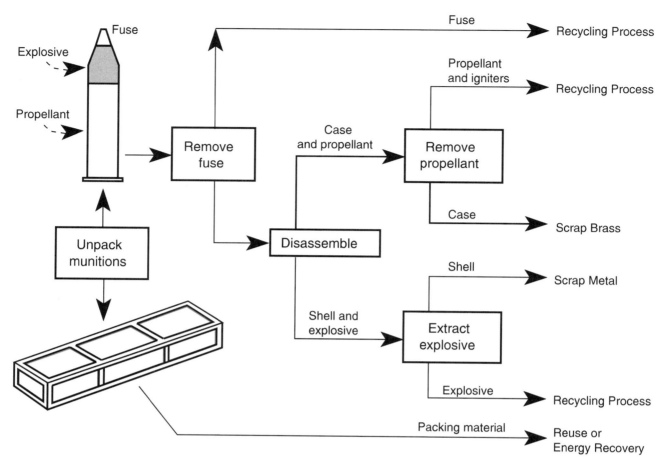

FIGURE 3-11. Munition disassembly steps (adapted from Hermann [42])

chemical filling ingredients generally are not re-covered due to the variability of the composition used, their sensitivity, and the low value of their ingredients (43).

3.11.5 Advantages

Extraction is a necessary preprocessing step for most options to reuse or recover energetic materials. With the exception of devices to be refurbished and reused, the energetic material first must be removed from the device to allow additional processing.

3.11.6 Disadvantages and Limitations

Due to the nature of the available energy content and low activation energy of the materials, all pro-cessing of energetic material requires careful atten-tion to safety precautions to avoid initiation of high-energy release events.

Sludges and soils containing less than 10 percent by weight of energetic materials typically pass the U.S. Army Environmental Center criteria for non-reactivity and do not exhibit a RCRA ignitability or reactivity characteristic. Soils containing higher concentrations require special precautions. Ener-getic materials extracted from sludges or soils are likely to be sufficiently concentrated to require spe-cial precautions (43).

Explosives projectiles and the oxidizer and fuel in rocket motors are held by a binder, which usually is a crosslinked thermosetting polymer. The binder can complicate solvent extraction of explosives or the aqueous dissolution of water-soluble oxidizers (44).

3.11.7 Operation

For an energetic material to be recycled, it typi-cally must be removed from its current container (e.g., projectile body or rocket motor casing). Conventional techniques involve some combination of disassembly and punching or cutting to gain access to the energetic material.

Munition components can be disassembled and separated by a process called reverse engineering, which involves separation of the casing (containing ignition compounds and propellant) from the projectile ignition compounds, explosives, and possibly a fuse. The propellants are easily removed from metal casings, allowing both the energetic materials and metals in the casing to be reused. Projectiles or bombs can be opened by a variety of methods. Punching opens small items with thin- or medium-thick walls, such as pyrotechnic or smoke munitions. Shearing with a guillotine-like shear blade removes fuses and cuts rocket motors into smaller sections. Wet saw cutting or high-pressure water jet cutting are applicable to a wide variety of munition types (45). Equipment for reverse engineering can be designed to work well for specific munitions but does not adapt easily to varying configurations (43).

Once the container is opened, the energetic material can be removed. For composite rocket motors and other items containing energetic materials held in place by binders, high-pressure water washout (hydromining) and mechanical cutting (machining) are the established methods to remove the energetic materials from the container. Hydromining has been in commercial operation since the mid-1960s to remove energetic materials from rocket motors and projectile bodies. Propellant machining is a standard manufacturing technique that shapes the initial burning surface in a rocket motor to provide the required ballistics or to remove all of the propellant from rocket casings (43).

Cryogenic washout is a dry process that uses high-pressure jets of cryogenic liquid to embrittle and fracture the energetic material. Bench-scale testing has been performed with liquid nitrogen and liquid ammonia, and large-scale tests are planned (43). Removal of energetic material using CO_2 pellet abrasion and critical fluid extraction also is under development (46).

Methods to dissolve the polymer binders used to hold energetic materials also are being developed. Polyurethane-based polymers are commonly used as binders for propellants and explosives. By undergoing hydrolysis at 230°C (445°F), the polyurethane groups in the binder split. The mixture is then treated by solvent extraction to recover both polyols and energetic materials from the binder (47). (For more information on the use of chemolysis to reduce polymers to monomers and oligomers, see Section 3.15.)

Some munition binders are heat sensitive and degrade upon heating. Polypropylene-glycol-urethane, for example, will degrade when heated to 160°C (320°F) and held for 10 hours (44).

Melting and steamout are well-established methods for removing TNT from explosive devices. These processes use heating to liquefy the TNT, which is then poured out of the casing. Melting and steamout are in commercial-scale use at a variety of ammunition plants and at the U.S. Army's Western Demilitarization Facility in Hawthorne, Nevada (43).

An emerging technique uses fracturing at cryogenic temperatures to open the container and extract energetic materials. Cryogenic fracturing involves cooling the device with liquid nitrogen followed by crushing in a hydraulic press (48). Both the metal casing and the energetic fill are brittle at cryogenic temperatures, so the device fractures into small pieces when crushed. The fragments can be processed to recover the energetic materials by solvent extraction, melting, gravity separation, or magnetic separation (43).

Solvent extraction is the most appropriate process for recovery of water-insoluble explosives from contaminated soils, sludges, and process wastes. Washing explosives-contaminated lagoon samples with a 90-percent acetone and 10-percent water extractant has been demonstrated to achieve greater than 99-percent removal. Recovery of the explosives and regeneration of the extractant, however, present significant challenges. Distillation is the only currently feasible method for separating the extracted explosive from the acetone/water solvent. The distillation process subjects the acetone to elevated pressure and temperature. Exposing a volatile solvent containing the extracted explosives to distillation conditions raises serious safety concerns. An alternative solvent regeneration method would be needed to allow commercial-scale development of a solvent extraction system for wastes contaminated with explosives (43).

3.12 PROPELLANT AND EXPLOSIVE REUSE

3.12.1 Usefulness

Physical and chemical methods are available to reuse energetic materials in similar applications.

3.12.2 Process Description

Obsolete munitions and rocket motors can be inspected and reused for training or similar applications. Explosives and energetic materials can be remanufactured into new explosive products, or processed to separate and recover the energetic material for reuse (see Figure 3-12).

3.12.3 Process Maturity

Munitions and rocket motors have been inspected and reused on a limited scale. Remanufacture of new devices from obsolete equipment has been demonstrated on a small scale, and reuse of separated energetic material has been demonstrated on a commercial scale (43).

3.12.4 Description of Applicable Wastes

Relatively stable high explosives such as high-melting explosive (HMX, or octahydro-1,3,5,7-tetranitro-1,3,5,7-tetrazocine), 2,4,6-tetranitro-N-methylaniline (tetryl), or TNT can be reliably reclaimed and reused. Propellants such as nitrocellulose (NC), dinitrotoluene (DNT), dibutyl phthalate (DBP), and nitroglycerine (NG) and oxidizers such as ammonium perchlorate (AP) are less stable and may require significant purification prior to reuse (44).

3.12.5 Advantages

Reuse of energetic materials allows potential waste material to be recovered as a high-value product and avoids the necessity of using new resources to manufacture explosives.

3.12.6 Disadvantages and Limitations

Due to the nature of the available energy content and low activation energy of the materials processed, all processing of energetic materials requires careful attention to safety precautions to avoid initiation of high-energy release events.

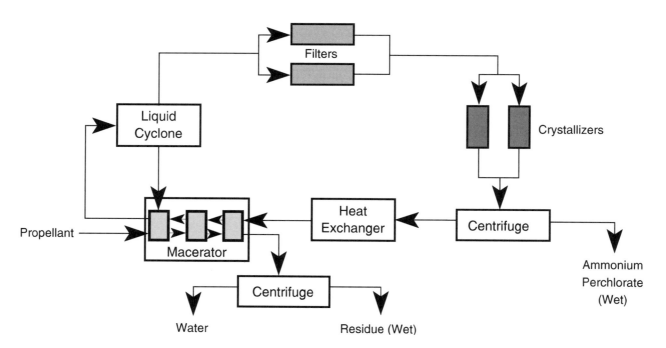

FIGURE 3-12. The ammonium perchlorate reclamation process (43)

3.12.7 Operation

Ordnance items and rockets are routinely rein-spected for training and similar applications. Reuse is, of course, applicable only to devices that are in good condition and have a well-documented history.

Hazard class 1.1 rocket propellant, containing explosives such as NG, NC, and HMX, has been remanufactured into 2-lb booster charges used to initiate ammonium nitrate/fuel oil or slurry explosives. Plastic-bonded explosives have been granulated and reused to make charges for metal bonding and forming applications (43).

Energetic compounds can be collected for reuse by processing to reject binder, impurities, and other inert components. Explosives such as high-blast explosive (HBX), HMX, research department explosive (RDX, or hexahydro-1,3,5-trinitro-1,3,5-triazine), tetryl, TNT, NG, and NC are dissolved or suspended by steaming, high-pressure water jet cutting, or solvent extraction (see Section 3.6). Filtration, selective extraction/precipitation, vacuum evaporation, and other purification methods then separate the explosives from the binders and impurities, such as metal fragments and decomposition products (44).

Purified surplus explosive can undergo large-scale commercial reuse in slurry explosives. Slurry explosives are a saturated aqueous solution of water-soluble oxidizer, which carries particles of oxidizer and sensitizing "fuel" in suspension. The most common oxidizer is ammonium nitrate, and the most common sensitizer is aluminum powder (49). Sodium nitrate, sodium perchlorate, and sodium chlorate are possible alternative oxidizers. Patent literature shows that munition explosives such as TNT, tetryl, HMX, RDX, and NG are used as sensitizers in slurry explosives. The reported consumption of slurry explosives is hundreds of millions of pounds annually (43).

Water-soluble ammonium perchlorate is recovered from composite rocket propellants by leaching with hot water. The propellant mixture, consisting of binder, ammonium perchlorate, and aluminum, is size reduced and contacted with heated water in a macerator. The ammonium perchlorate is recovered from the water by crystallization. The recovered ammonium perchlorate is indistinguishable from salt made from new materials and can be reincorporated into rocket propellant (43).

3.13 PROPELLANT AND EXPLOSIVE CONVERSION TO BASIC CHEMICALS

3.13.1 Usefulness

Chemical processing is used to convert propellants and explosives to basic chemicals that can be reused.

3.13.2 Process Description

The energetic components of munitions may have commercial use as basic chemicals rather than as explosives (see Figure 3-13).

3.13.3 Process Maturity

Commercial processes are available to recover basic chemicals from munitions. Applications have been limited to a few special situations due to the low value of the basic ingredients.

3.13.4 Description of Applicable Wastes

Energetic materials that contain a high proportion of ammonia or nitrate are potentially useful for fertilizer manufacture. Materials such as aluminum in rocket propellants or zinc, manganese, and phos-

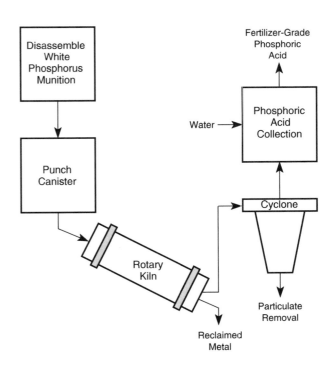

FIGURE 3-13. *The white phosphorus reclamation process*

phorus in pyrotechnic or smoke munitions can be recovered.

3.13.5 Advantages

Conversion to basic chemicals can open a wider market for lower value energetic or pyrotechnic materials.

3.13.6 Disadvantages and Limitations

Due to the nature of the available energy content and low activation energy of the materials processed, all processing of energetic and pyrotechnic materials requires careful attention to safety precautions to avoid initiation of high-energy release events.

3.13.7 Operation

Ammonium nitrate and ammonium perchlorate-based propellants can be ground to reduce the particle size, blended with inert carriers, and reused as nitrogen fertilizer (50). Purified nitrocellulose-based propellants can be used as supplements in animal feed (46).

The Crane Army Ammunition Activity in Crane, Indiana, recovers white phosphorus from munitions by converting the phosphorus to phosphoric acid. The process produces marketable phosphoric acid and metal scrap. The acid conversion plant processes munitions from other Army facilities and has sold thousands of tons of phosphoric acid and scrap metal for its demilitarization operations (43).

Thermolysis (Section 3.17) using a hydrogenation process is being developed to convert propellants and energetic materials to useful chemicals. In this process, the waste is combined with hydrogen and is heated over a catalyst in the temperature range of 40°C to 400°C (100°F to 750°F) at pressures ranging from 1.6 to 8.6 mPa (250 to 1,250 psi) to form recoverable light organic chemicals, such as methane and ethane (51).

3.14 RE-EXTRUSION OF THERMOPLASTICS

3.14.1 Usefulness

Reprocessing of thermoplastic waste is a method used to make commercial and industrial polymeric products from postconsumer and postindustrial commingled plastics. The material is mechanically reground and then extruded into the required shapes. (Sources for information on specifications for thermoplastic material reuse are given in Section 4.3.)

3.14.2 Process Description

The various stages for recycling thermoplastic wastes are polymer characterization, collection, separation, cleaning, regrinding, and extrusion. Postconsumer thermoplastic products typically are made with extrusion molds, blow molds, or injection molds (see Figure 3-14). Extrusion entails rotating a screw in a barrel to melt plastic pellets and force the molten resin through a die. Extrusion usually precedes blow/injection molding. The extrusion process is used to form film plastics (such as sheet wraps) or profile extrusions (such as pipe). Blow molding is used for containers (such as bottles), whereas injection molding is used to form solid parts (such as bottle caps) that require higher dimensional precision.

One common structural product made from commingled (mixed) plastic waste streams is plastic lumber, which is a flow-molded linear profile. A mixture of films and containers, as well as some residual impurities known as tailings, are blended into a compatible raw material and extruded into large cross-section items that have structural utility. Blending is enhanced by compatibilizers that allow bonding between two otherwise unadhering plastics. Other types of impurities are "encapsulated" during the extrusion of these shapes. Recently, wood, flour, and glass fibers have been mixed with the recyclate to enhance the mechanical properties (such as stiffness and strength) of the lumber. Production of plastic lumber is the main focus of this discussion, as plastic lumber is most amenable to the use of mixed plastics (53).

3.14.3 Process Maturity

Research currently is under way to improve the characterization and separation of various polymers in waste streams. The technology for regrinding and extruding products itself is mature. The extruders and molding equipment that process recyclates are commercially available. Both structural and non-load-bearing products made from recycled plastics are in the marketplace. Only a few companies, how-

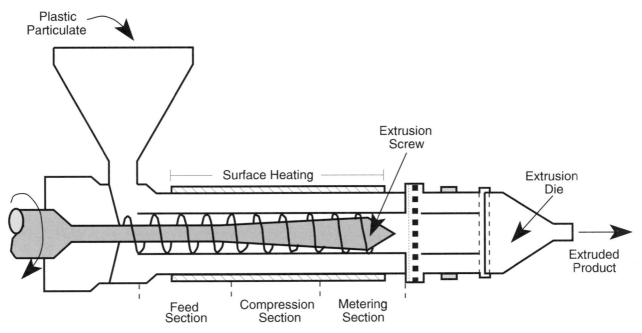

FIGURE 3-14. Example of an extruder (52)

ever, currently make glass- and fiber-reinforced plastic lumber with proprietary technologies.

3.14.4 Description of Applicable Wastes

The six types of plastics currently identified by number for recycling are polyethylene-terephthalate (PET), high-density polyethylene (HDPE), polyvinyl chloride (PVC), low-density polyethylene (LDPE), polypropylene (PP), and polystyrene (PS). All other plastics are included in Category 7 for recycling purposes. The polyethylenes (PEs), PP, and PS seem to be most suitable for plastic lumber production. Some compatibilizers currently are available to extrude PE/PVC, PE/PS, and PVC/PS blends, as well as some specialized plastics. The blends can be used to extrude plastic lumber. PVC in large fraction does lead to some stability problems due to degradation (52, 54, 55).

3.14.5 Advantages

Mechanical recycling or re-extrusion of thermoplastic waste has several advantages in addition to reducing waste disposal volume. First, commingled plastics can be processed with minimal separation. Next, manufactured plastic lumber has weather-re-

sistant properties superior to those of traditional wood, as well as sufficient strength and toughness to replace wood. Furthermore, the re-extrusion process is a way to use tailings, the miscellaneous plastics left after the stream has been mined of HDPE and PET. Finally, certain products (such as plastic pallets) can be molded directly rather than fabricated from extruded lumber.

3.14.6 Disadvantages and Limitations

The user of materials from Superfund sites will have a high level of concern about the potential for incorporating trace contaminants into commercial products. The primary limitation of re-extrusion of thermoplastic waste is the variability in the properties of the end product. In addition, very few specifications for lumber products currently are available to the buyer to purchase these products. ASTM Committee D20.95 currently is working on some of these standards (see Section 4.3). Another limitation of the lumber seems to be lower stiffness values compared with wood that sometimes lead to unacceptable deflection in structural uses. Therefore, changes in design sometimes are warranted for structural applications. Several research programs funded by state and federal agencies currently are under way to establish material property databases and design procedures.

3.14.7 Operation

Machinery for extruding and molding currently is available on a turnkey basis and can be set up for operation. A directory of equipment manufacturers with a list of products made from recycled plastics has been compiled. The availability and transportation costs of the raw materials heavily influence the economics of a successful recycling plant, however. Extrusion without separation leads to products of dark brown, black, or gray colors; for other colors, separation and sorting are necessary. A manufacturer may require that a municipality that collects mixed plastics also buy the recycled product. Studies have shown that local, state, and federal governments will be the largest consumers of recycled plastic lumber products, such as guardrail posts, plastic pallets, benches, or landscape timbers.

3.15 CHEMOLYSIS

3.15.1 Usefulness

Chemolysis is a chemical method to recover useable monomers or short-chain polymers from solid polymer wastes.

3.15.2 Process Description

Chemolysis is a depolymerization reaction to convert polymers (about 150 repeating units) into monomers or short-chain oligomers (2 to 10 repeating units). Chemolysis reduces condensed polymers by reversing the preparative chemistry (see Figure 3-15). The polymers react with water or alcohols at elevated temperature to break the bonds between units (56).

3.15.3 Process Maturity

Chemolysis is used commercially to recycle clean waste that contains one type of polymer. In particular, several companies have commercial processes that recycle PET wastes by methanolysis, glycolysis, or hydrolysis. Extension of the process to mixed polymer types will require significant development (57). Molecular separation of polymers by selective dissolution has been demonstrated at the bench scale (58).

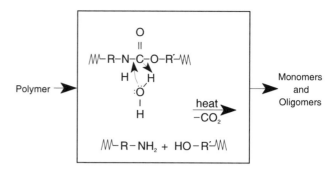

FIGURE 3-15. Example of a chemolysis reaction

3.15.4 Description of Applicable Wastes

Condensation polymers such as polyesters and poly-amides and some multiple addition polymers such as PEs are created by reversible reactions; therefore, chemical reaction can convert them back to their immediate precursors. Bulk thermoplastics such as HDPE, PP, PS, and PVC are not easily broken down by chemical methods. These bulk plastics are more amenable to thermal treatment to produce hydrocarbon products (see Section 3.17) (59, 60).

3.15.5 Advantages

Converting a solid plastic back into the monomers used for manufacture creates a high-value chemical product. Unlike grind and remelt recycling, conversion to monomers and oligomers allows purification and remanufacture of the plastic. Remanufacture avoids problems with impurities such as copolymers, stabilizers, and pigments, and with heat history and aging effects such as yellowing or embrittlement (61).

3.15.6 Disadvantages and Limitations

The reaction conditions are specific to one type of polymer and generally do not work well with mixed plastic types. Chemical recovery of organic chemicals from plastics requires expensive equipment and therefore high throughput for economical operation.

3.15.7 Operation

Using glycols as the reaction medium stops short of complete conversion to monomers. Glycolysis produces short-chain oligomers consisting of 2 to 10 units. Use of excess methanol at elevated temperature as the reaction medium results in conversion to monomers. For example, PET can be heated in contact with methanol to produce dimethlyterephthalate and ethylene glycol (62). Recent European patent applications discuss treatment of PET wastes with alcohol and a barium hydroxide transesterification catalyst to produce soluble polyesters. Urethane can be broken down with an alkanolamine and catalyst into a concentrated emulsion of carbamates, ureas, amines, and polyol (59).

Selective dissolution can also separate mixed polymers. In this process, single polymers are removed from a mixture by dissolution at a controlled temperature. The most soluble polymer is removed first, followed by sequential dissolution of polymers at increasing temperatures (52, 58).

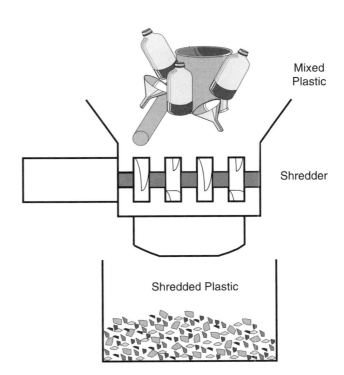

FIGURE 3-16. Example of a plastic shredding operation

3.16 SIZE REDUCTION AND REUTILIZATION OF PLASTIC AND RUBBER WASTES

3.16.1 Usefulness

Size reduction and reuse of plastics and rubber wastes are ways in which waste plastics and rubber are shredded or ground for use as filter media, filler in new plastics, concrete or asphalt aggregate, and other applications requiring low-density filler. General requirements for rubber particulate for reuse are described in Section 4.4.

3.16.2 Process Description

Originally, emphasis in plastics recycling was placed almost entirely on thermoplastic materials because they could be easily reworked by melt processing (see Section 3.14). Thermoplastic materials accounted for the largest volume of plastics in packaging and durable goods, encompassing such commodity plastics as PE, PP, PS, PVC, and PET polyesters. Thermoset plastic materials have received greater emphasis in recent years as attention has shifted from packaging to durable goods, and recycling of thermoset plastics is now feasible. Thermoset materials can be shredded or ground for reuse (see Figure 3-16). The focus of recent research is to allow the plastic particulate to be reclaimed for higher-value applications.

3.16.3 Process Maturity

Clean, single-type thermoset plastic particulate is reused commercially as filler in new plastic products. Applications for mixed materials require further development. Mixed plastic is most likely to be reused as aggregate or as a supplement in cement. The preparation, characterization, and testing of polymer concrete (PC) and polymer mortar (PM) are in the research phase.

The use of scrap tire in asphalt pavements is also in the research stage. Although commercial equipment is available to grind and separate rubber from scrap tires and mix it with asphalt, the impact of mixing on pavement performance is not yet certain. Several state governments and the Federal Highway Administration (FHWA) have conducted feasibility studies and trial tests. All of these studies and tests indicated that this method is the most effective way to reuse scrap tires. The effect of rubber in asphalt on pavement performance is still being evaluated, and a cost-benefit analysis of this new construction material's use on a large scale is being conducted. On a more limited scale, recreational and sport surfaces prepared from this material have worked quite successfully (54).

3.16.4 Description of Applicable Wastes

Lower-value mixed thermoplastic materials or thermosetting materials can be ground and reused as particulate. (Extrusion of waste thermoplastics is discussed in Section 3.14.) Common thermoset polymers include vulcanized elastomers (both natural and synthetic rubbers), epoxies, phenolics, amino resins (e.g., urea and melamine formaldehyde), polyurethanes, and polyesters (based on unsaturated polyester resins, generally modified and crosslinked with styrenic monomers).

3.16.5 Advantages

The main advantage of using waste plastics in PC and scrap tires in asphalt is that both PC and asphalt consume large quantities of these solid wastes, thereby reducing waste disposal and alleviating incineration concerns. Other advantages include enhancement in the strength-to-weight ratio of concrete and durability compared with cement-based materials.

3.16.6 Disadvantages and Limitations

One of the main disadvantages of PC and PM is the loss of strength at high temperatures, which is an important factor in applications such as precast building panels. For rubberized asphalt, the problem so far has been justifying the increased cost of the material, as the benefits of better performance and durability are not yet established.

3.16.7 Operation

Crosslinked polyurethanes are processed by shredding or grinding for reuse in reaction injection molding (RIM) or in plastic foam as inert filler or "rebond." Blending flexible foam crumb waste with 10 to 20 percent by weight virgin liquid isocyanate/polyol prepolymer and using the reacted, cast composition as carpet backing has been commercially successful. On the other hand, attempts to use polyurethane regrind waste in RIM systems have been limited to low regrind levels (on the order of 10 percent by weight), because the viscosity of the liquid components that carry the regrind quickly reaches unacceptable levels for the RIM process.

Recycling of sheet molding compounds (SMCs) and bulk molding compounds (BMCs) differs from polyurethane recycling in two important respects:

1) the polyester/styrene bonds are not thermally reversible below the degradation temperature of the polymer and 2) SMCs and BMCs have a very high inorganic filler loading (on the order of 70 percent by weight). This high filler loading forestalls the use of incineration as a disposal method because the recoverable energy content per unit weight is low, the heat sink burden is high, and the volume of residue is high.

Early recycling attempts centered on grinding the crosslinked, filled compounds to a powder for use as an inert filler. Ground SMC filler could best be incorporated into the SMC mix by withholding an equivalent amount of calcium carbonate, resulting in a compound that had the advantage of a slightly lower density. The monetary value of the calcium carbonate replaced by ground SMC, however, is a few cents per pound and cannot account for the processing costs of grinding SMC.

To upgrade the value of recycled SMC, processes were developed that enable the composite to be broken up gently, leaving the glass fibers largely intact. SMC or BMC products prepared with this type of recycled polyester composite benefit from the reinforcement provided by the salvaged glass fibers. These products can replace a portion of both high-quality, primary calcium carbonate and the glass fibers required for compounding new SMC or BMC materials. Processes along similar lines are being developed.

Plastic or elastomer particulate also can be reused in concrete or asphalt. PET and PE chips are mixed with concrete or mortar to form PC or PM. These composites are stronger and more durable than conventional cement-based construction materials (63). Rubber from old tires can be reused in pavement applications to enhance pavement performance. Two processes yield this composite material; the "dry" process uses 6.35 mm (¼ in.) chunks or larger and the "wet" process uses finely ground particles of rubber (64).

3.17 THERMOLYSIS

3.17.1 Usefulness

Thermolysis (thermal conversion) uses elevated temperature in a controlled atmosphere to produce valuable chemical monomers by processing liquid or solid organic wastes.

3.17.2 Process Description

Thermolysis of high-organic solid wastes to basic hydrocarbon products involves pyrolysis, hydrogenation, or gasification. Pyrolysis is heating in the absence of air to produce liquid and gaseous hydrocarbons (see Figure 3-17). Hydrogenation is the treatment of viscous organics at high temperature and pressure, typically with a catalyst, to produce valuable saturated hydrocarbons. Gasification is partial oxidation of a range of hydrocarbons to produce synthesis gas (carbon monoxide and hydrogen), used in the production of organic chemicals (56, 57).

3.17.3 Process Maturity

In Europe, plastics pyrolysis demonstration plants are operating in Ebenhausen, Germany, and Linz, Austria. Development in the United States to date has focused on treatment of scrap tires. The current development challenges are process scale-up and improved process efficiency when processing high-solids-content plastics (65).

3.17.4 Description of Applicable Wastes

Thermochemical methods to convert organic solids to petrochemicals are better able to process mixed material and higher inorganic content than re-extrusion or chemolysis are. Mixed polymeric wastes can be processed without sorting or significant preparation. Contamination of the plastic wastes also can be tolerated (66).

3.17.5 Advantages

Thermolysis can treat mixed, coarsely ground plastic scrap at high throughputs to yield basic hydrocarbon products. The process can be designed and operated to limit the production of higher molecular-weight liquids that can reduce overall process efficiencies.

3.17.6 Disadvantages and Limitations

Chlorine (from PVC wastes) contaminates the liquid product if not removed. Even low levels of chlorine are unacceptable for refinery feed or fuel use. Testing of a thermolysis process at the Energy and Environmental Research Center (EERC) at the University of North Dakota targeted a maximum chlorine content of 200 ppm (67). Chlorine can be removed by preprocessing or by retention in fluidized bed material such as calcium oxide (CaO).

Some nitrogen-containing polymers, such as nylon, can produce hydrogen cyanide, resulting in

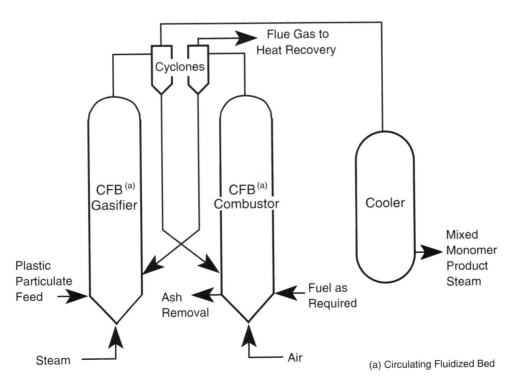

FIGURE 3-17. Example of the pyrolysis process

potentially hazardous off-gas. The process uses high-temperature processing equipment, which requires a significant capital investment to develop a full-scale processing plant.

3.17.7 Operation

Testing at the EERC indicates that thermal decomposition of polymers is influenced by temperature, fluidized bed material, fluidization velocity, and feed polymer composition. Temperature had the strongest effect, but the effects of bed material, fluidization velocity, and feed composition were greater than expected. With olefin polymers, increasing temperature changed the product liquid composition; lower temperatures yielded olefins and aliphatics, intermediate temperatures yielded cyclics and aromatics, and high temperatures yielded fused-ring aromatic polymers (67).

A typical process uses circulating fluidized bed reactors to convert mixed plastic wastes into monomer feedstock. The primary product from the process is ethylene, based on the composition of typical plastic wastes. A product gas containing about 40 percent ethylene has been produced from a mixed polymer feedstock. The product gas could be fed to an ethylene purification plant to produce the high-purity feedstock necessary for polymerization and/or other products. Treatment of coarsely ground plastic scrap at high throughputs produces the desired monomer products at low cost. Operation of the process is expected to be simple, with a startup fuel used initially to heat the reactors and supply any additional energy required for the system. Severe operating conditions (e.g., high temperature and high pressure) allow conversion of industrial wastes, scrap tires, used oil, or mixed plastic wastes to basic organic components (68).

3.18 CHEMICAL PRECIPITATION

3.18.1 Usefulness

Precipitation is a chemical method to remove and concentrate dissolved inorganics from aqueous materials with low concentrations of contaminants. The precipitated solid may be a useful product, or it may receive additional processing (e.g., chemical leaching or smelting) to recover a salt or metal.

3.18.2 Process Description

In the chemical precipitation process, soluble contaminants are removed from a waste stream by their conversion to insoluble substances. Physical methods such as sedimentation and filtration can then remove these precipitated solids from solution. The addition of chemical reagents that alter the physical state of dissolved or suspended metals initiates precipitation. Reagents also adjust the pH to a point where the metal is near its minimum solubility (see Figure 3-18). Standard reagents include:

- Lime (calcium hydroxide)
- Caustic (sodium hydroxide)
- Magnesium hydroxide
- Soda ash (sodium carbonate)
- Trisodium phosphate
- Sodium sulfide
- Ferrous sulfide

These reagents precipitate metals as hydroxides, carbonates, phosphates, and sulfides. Metals commonly removed from solution by precipitation include arsenic, barium, cadmium, chromium, copper, lead, mercury, nickel, selenium, silver, thallium, and zinc. Xanthates also are being evaluated for the precipitation of metals (69).

The chosen reagent is added to the metal solution by rapid mixing followed by slow mixing to allow the precipitate particles to grow and/or flocculate. A flocculant/coagulant reagent often is needed to enhance particle agglomeration. Sedimentation or clarification, the next step in precipitation, allows the precipitate to settle to the bottom of a tank for collection. Typical settling times for heavy metal particles range between 90 and 150 min (70).

3.18.3 Process Maturity

Industrial use of chemical precipitation of metal-containing wastewater is widespread. Precipitation is commonly used for wastewater treatment at electroplating facilities, leather tanning shops, electronics industries, and nonferrous metal production facilities. Precipitation often is selected for removing metals from ground-water pump-and-treat operations. Vendor systems are available for ground-water treatment with design flowrates of up to 95 L/sec (1,500 gal/min).

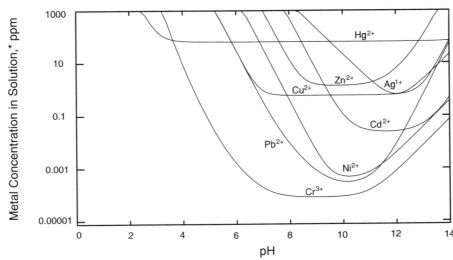

FIGURE 3-18.
Solubility of metal ions in equilibrium with a hydroxide precipitate.

Precipitation also is applied to recycling, either to capture metals in a more concentrated form for subsequent processing or, less commonly, to form a solid metal salt product.

3.18.4 Description of Applicable Wastes

Hydroxide precipitation effectively removes cadmium, chromium (+3), copper (+2), iron (+3), lead, manganese, nickel, tin (+2), and zinc. Most of these can be removed to concentrations of less than 1 mg/L (0.06 grains/gal). For arsenic, cadmium, lead, silver, and zinc, the residual concentration in solution is significantly lower with sulfide precipitation. Sulfide precipitation also is effective for chromium (+3, +6), copper (+1, +2), iron (+2, +3), mercury, nickel, and tin (+2). Hydroxide precipitation generally shows poor metal removal when chelating agents are present, although these agents have less effect on the efficiency of sulfide precipitation.

3.18.5 Advantages

Precipitation collects metals in a concentrated form for reuse as salts or, with additional processing, for metal recovery. Precipitation can achieve very low concentrations of metal contaminants in the treated water. Chemical precipitation is a proven technology for industrial wastewater treatment with many years of data demonstrating effectiveness for a wide range of waste streams. Precipitation uses simple, low-capital-cost, commercially available equipment.

3.18.6 Disadvantages and Limitations

Precipitation usually does not produce a useful product directly. In most applications, the precipitated sludge requires further processing to recover the metal. For example, precipitation may be used to produce a sludge that is then processed by hydrometallurgical or pyrometallurgical methods to recover products. Precipitation usually requires supplementary processing or chemical adjustments. For example, physical separation may be needed to remove suspended solids or oil and grease. Adjustment of the metal's oxidation state (e.g., reduction of Cr(VI) to Cr(III)) may be needed.

Some precipitation processes, particularly the sulfide system, have the potential to generate undesirable or toxic sludge or off-gas.

3.18.7 Operation

A typical system may begin with pretreatment steps such as large solids removal, cyanide destruction, and/or chemical reduction. Materials then enter a mixing tank, where the selected reagent is added. A flocculation/coagulation tank may then be needed to allow the precipitate to settle. Flocculants such as alum, lime, or polyelectrolytes are added in a slowly stirred tank to promote agglomeration of the precipitate, yielding denser particles that settle faster. Settling usually takes place in a clarifier with a sloped bottom, where the sludge is collected. In a continuous system, the sludge is sent to a subsequent settling tank for further settling. Following settling, the sludge must be dewatered through vacuum filtration or filter presses prior to recovery.

Treated water effluent from the clarifier can be filtered to remove fine particulates that did not settle. Other technologies such as ultrafiltration, reverse osmosis, activated carbon capture, and ion exchange can be used to treat the effluent to further reduce concentrations of metals, if required.

The pH is a critical operating parameter in a chemical precipitation system. The optimal pH must be first determined, then maintained. A pH level at which the metal compound has a low or minimum solubility is desired. For hydroxides, the pH would range from 9 to 11. Sulfides have lower solubilities at similar pH ranges; therefore, lower pH levels (pH 8 to 10) can be used to achieve comparable removal. Selecting the optimal pH is complicated by the fact that metal hydroxides have specific points of minimum solubility, and these minimums occur at different points for different metals. Varying solubilities present a challenge when designing a system to treat a water stream containing several metals. Including more than one precipitation stage with different pH points may be necessary. Sulfide precipitation normally is done at a pH of greater than 8 to avoid generation of hydrogen sulfide gas. Other operating parameters include retention time in various process steps, flowrate of the waste stream, reagent dose rates, and temperature.

3.19 ION EXCHANGE

3.19.1 Usefulness

Ion exchange is a physical/chemical method to concentrate and recover dissolved inorganics from aqueous solutions with low concentrations of contaminants.

3.19.2 Process Description

Ion exchange is a technology to remove ionic species, principally metals ions, from aqueous waste streams. The process is based on the use of specifically formulated resins having an "exchangeable" ion bound to the resin with a weak ionic bond. If the electrochemical potential of the ion to be recovered (contaminant) is greater than that of the "exchangeable" ion, the exchange ion goes into solution, and the metal binds to the resin (see Figure 3-19).

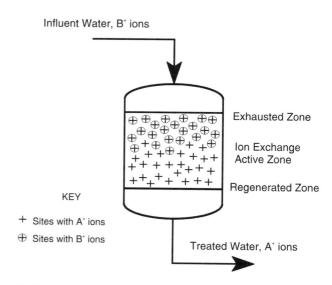

FIGURE 3-19. Example of an ion exchange operation

Resins are separated into two classes: cation and anion. Cation resins exchange positive ions, such as dissolved metals, and anion resins exchange negative ions, such as sulfate or nitrate. Resins have a higher affinity for some ions than for others. Generally speaking, strong-acid resins prefer cations with higher ionic charges.

Ion exchange is reversible, so the captured metal ions are removed from the resin by regeneration using an acid for cation resins or a base for anion resins. The concentration of contaminants is higher in the regeneration solution than in the treated wastewater. The regeneration solution is further treated to recover metals or salts.

3.19.3 Process Maturity

Ion exchange technology is fully developed and commercially available, although applications are waste-stream specific. Applications have included removal of radionuclides from power plant waste streams, removal and recovery of metals from electroplating operations, removal of metals from ground and surface waters, recovery and removal of chromium, and deionizing/softening of process water.

3.19.4 Description of Applicable Wastes

Ion exchange systems can effectively remove ionic metals such as barium, cadmium, copper, lead, mercury, nickel, selenium, silver, uranium, and zinc. The technology also is applicable to nonmetallic anions such as halides, and to water-soluble organic ions such as carboxylics, sulfonics, and some amines. Ion exchange has been used to recover chromium and copper from ground water contaminated with wood-treating chemicals. The chromium and copper are returned to the wood-treating plant for reuse (71). Ion exchange in general is applicable for recovering metals from solutions containing less than 200 mg/L (12 grains/gal) of dissolved metal. Solutions with higher concentrations usually are more efficiently concentrated using electrowinning (Section 3.28) or dialysis methods (Sections 3.22 and 3.23) (70).

3.19.5 Advantages

Ion exchange is capable of extracting all metals from dilute wastewater streams and collecting the metals, at a much higher concentration, in the regeneration solution. Further processing usually is needed to produce a product metal or salt, however.

The capital cost of ion exchange equipment is low, and operating costs are most influenced by chemical use, cost of resin, and labor for regeneration. The cost to treat a fixed water flow increases as the dissolved ion concentration increases. Therefore, ion exchange is best suited for removal of metals from waste streams with lower concentrations of dissolved metals (2).

3.19.6 Disadvantages and Limitations

The ion exchange resin must be regenerated to remove the metals collected from the wastewater. Most ion exchange processes are not selective and thus recover a mixture of metals. A reusable product is not recovered directly; the regeneration solution is further processed to produce a reusable product. A technology such as dialysis or electrowinning is used to recover metals or salts from the concentrated regeneration solution.

Ion exchange resins are prone to fouling because of high concentrations of suspended solids or some organic substances. Pretreatment such as filtration can remove some fouling agents. Oxidizers such as chromic or nitric acid may react with the ion exchange resin, causing the resin to degrade (70). In some cases, the reaction may be severe enough to result in serious safety concerns.

3.19.7 Operation

Ion exchange systems can operate in one of four modes: batch, fixed bed, fluidized bed, and continuous. The fixed bed system, the most common, typically includes four steps: service, backwash, regeneration, and rinse. During service, the waste stream is passed through the ion exchange resin, where ionic contaminants are adsorbed. When the ion exchange resin is nearing full capacity (the concentration of dissolved ions in the effluent begins to increase), the ion exchanger is taken out of service and backwashed to remove suspended solids. The resin bed is then regenerated by passing either an acid (for cationic resins) or a caustic (for anionic resins) through the resin bed. The resultant low-volume regeneration solution can then be further processed to recover metals or salts. The resin is then rinsed of excess regenerant and returned to service (72).

Factors that affect the performance of an ion exchange system include the concentration and valence of the contaminants, the concentration of competing ionic species or interferences, the concentration of total dissolved and suspended solids, and the compatibility of the waste stream to the resin material. Pretreatment techniques for ion exchange systems (e.g., carbon adsorption, aeration, or filtration) may be necessary. Cartridge filters upstream of the resin bed can remove suspended solids. To prevent iron and manganese precipitation, pre-aeration followed by flocculation, settling, and filtration can be used (70).

3.20 LIQUID ION EXCHANGE

3.20.1 Usefulness

Liquid ion exchange (LIX) is a form of solvent extraction that allows separation, concentration, purification, and recovery of dissolved contaminants from solutions. The most common application is recovery of metals dissolved in water, but nonmetals and oil-soluble impurities also can be recovered.

3.20.2 Process Description

The basic principle of solvent extraction applied to recover metal from solution is simple. The process depends on shifting the reaction equilibrium of a system, usually by adjusting the pH of the aqueous phase. The organic extractant and operating conditions are selected to cause metals to partition from the waste stream into the extractant. A stripping solution is selected to favor distribution of the metals from the extractant into the stripping solution. Although real-world applications introduce complications, LIX is an effective method to treat wastewaters containing metals at low concentration while recovering a concentrated solution (73).

In the LIX process, an aqueous solution containing metal contaminants is contacted with an extractant. The extractant is immiscible in water and dissolved in high flashpoint kerosene. Typical extractants include dibutyl carbitol, organophosphates, methyl isobutyl ketone, tributyl phosphate, amines, and proprietary ion exchange fluids. The metal-containing water and the organic extractant phases are thoroughly mixed to allow rapid partitioning of the dissolved metals into the extractant (see Figure 3-20). After contacting, the mixed phases enter a settler, where the water and organic separate.

The organic extractant is then contacted with stripping solution in a second contacting/settling step. The stripping solution is chemically adjusted so that metals partition out of the organic extractant into the water. This stripping step regenerates the extractant for reuse and captures the metal in the aqueous stripping solution. The stripping water may be a marketable salt solution or may require further treatment by precipitation, electrowinning, or other processes to recover a salt or metal product (73).

3.20.3 Process Maturity

Large-scale commercial LIX operations for metal recovery have been in use for over 50 years. Equipment is available from several vendors, and a wide range of solvents allow extraction of the desired metals with minimum impurity levels.

The technology was developed initially for recovery and reprocessing of uranium. Applications have expanded to primary production of copper, vanadium, cobalt, rare earth metals, zinc, beryllium, indium, gallium, chromium, mercury, lead, iron, cadmium, thorium, lithium, gold, and palladium. Application to recovery of metals from waste streams is, however, a recent development (74, 75).

3.20.4 Description of Applicable Wastes

The technology can be used to recover a variety of dissolved metals at any concentrations. The total dissolved solid concentration can be any level. The technology currently is being demonstrated for dissolved metals concentrations ranging from 1 to 100,000 mg/L (0.06 to 6,000 grains/gal). The feed stream for an LIX process is a low-suspended-solids (preferably suspended-solids-free) aqueous fluid containing dissolved metal contaminants.

LIX can be used to regenerate extractant from chemical leaching (see Section 3.29). Metal is extracted from a metal-containing solid by aqueous solvents (acid, neutral, or alkaline solutions). LIX is then used to recover the metal and regenerate the extraction solution. LIX also can be used to recover metals from contaminated surface water, ground water, or aqueous process residuals.

LIX can selectively remove a dissolved metal contaminant from process brine, thereby allowing discharge or recycle of the brine. A metal or combination of metals can be recovered from multicomponent aqueous streams containing metals in chelated form or complexed by dissolved aqueous organics (e.g., citric acid) or inorganics (e.g., phosphates). Careful selection of extractant and process conditions is critical when applying LIX technology to

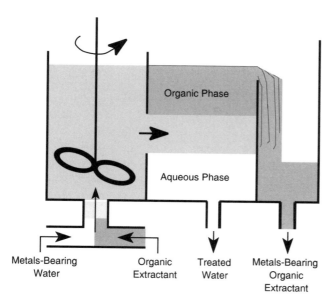

FIGURE 3-20. Liquid ion exchange contacting cell

these complex streams. Extractants also are available for recovery of toxic oxoanions, such as selenate and chromate.

3.20.5 Advantages

LIX processing recovers metals from solutions as high-purity, marketable compounds or elements. Extractants can be chosen to give high selectivity for the desired metal while rejecting common dissolved cations such as calcium or sodium and other impurities. Solutions containing as little as 1 mg/L to more than 100,000 mg/L (0.06 to 6,000 grains/gal) can be concentrated by factors of 20 to 200.

LIX can achieve high-throughput, continuous operation. The process is tolerant of variations in feed composition and flow. The extraction/stripping operations can be carried out in simple, readily available process equipment at low pressure and temperature (ambient to 80°C [175°F]).

3.20.6 Disadvantages and Limitations

The LIX process requires low levels of suspended solids. Suspended solids entering the phase separation settler may collect at the water/extractant interface, interfering with phase separation.

Complex waste streams can be processed but may require several extraction stages. Extractant formulation and process optimization for treating waste streams containing several metals and contaminants require experience with LIX processes and treatability testing.

3.20.7 Operation

Feed flowrates from 1 mL to thousands of gallons per minute can be processed. Various contactor/phase separation devices are available to allow selection of optimal equipment for the waste stream, extractant, and flowrate to be processed. For example, liquid-liquid contacting can be accomplished in mixer-settlers, columns, centrifuges, and hollow-fiber or spiral-wound membrane cells. After mixing in the contactor, the water/organic mixture is allowed to separate in a settler, a relatively quiescent area in the process equipment where the immiscible water and oil phases disengage and flow separately from the unit. The metals are stripped from the extractant using the same mixing/settling approach.

The stripping solution may be pure water or may contain acidic, basic, or neutral salts or dissolved complexing agents. The stripping chemistry is chosen to allow efficient removal of metal from the extractant while producing a useful product. For example, copper can be stripped by sulfuric acid to produce a marketable copper sulfate solution. The metal product from the LIX process may be a solution, solid salt, elemental metal, or precipitate. The stripping solution may be forwarded to a crystallizer or precipitator/clarifier to form solid salt or may be electrowon to form elemental metal.

3.21 REVERSE OSMOSIS

3.21.1 Usefulness

Reverse osmosis (RO) is a physical/chemical method to concentrate and recover dissolved inorganics in aqueous solutions with low concentrations of contaminants.

3.21.2 Process Description

Osmosis is the movement of a solvent (typically water) through a membrane that is impermeable to a solute (dissolved ions). The normal direction of solvent flow is from the more dilute to the more concentrated solution (see Figure 3-21). RO reverses the normal direction of flow by applying pressure on the concentrated solution. The semipermeable membrane acts as a filter to retain the ions and particles on the concentrate side while allowing water to pass through. The cleaned water passing through the membrane is called the permeate. The liquid containing the constituents that do not pass through

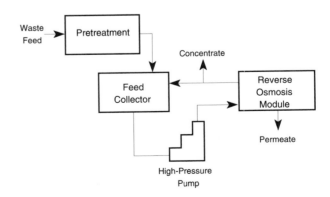

FIGURE 3-21. Example of the reverse osmosis process

the membrane (i.e., metals) is called the concentrate. Metal or salt products are recovered from the concentrate by techniques such as evaporation, electrowinning, or precipitation.

For RO applications, membranes that have high water permeability and low salt permeability are ideal. The three most commonly used RO membrane materials are cellulose acetate, aromatic polyamide, and thin-film composites, which consist of a thin film of a salt-rejecting membrane on the surface of a porous support polymer.

3.21.3 Process Maturity

RO has long been used to desalinate water. The process is beginning to be applied in the electro-plating industry for the recovery of plating chemicals in rinse water.

3.21.4 Description of Applicable Wastes

RO can be applied to concentrate most dissolved metal salts in aqueous solution. Electrolytes and water-soluble organics with molecular weights greater than 300 are stopped by the membrane and collect in the concentrate. Most metals in solution (e.g., nickel, copper, cadmium, zinc) can be concentrated to about 2 to 5 percent in the concentrate (76, 77). Waste solutions containing high suspended solids, high or low pH, oxidizers, or nonpolar organics typically are not suitable for RO processing.

3.21.5 Advantages

RO separates a waste stream into two streams: the cleaned permeate and the concentrate, which may contain percent concentrations of dissolved salts. The cleaned permeate can be reused as process water, and the concentrate can be further treated to recover metal or can be reused as a metal salt solution. RO is less energy intensive than distillation or evaporation.

3.21.6 Disadvantages and Limitations

RO membranes are sensitive to fouling and degradation. Even trace concentrations of nonpolar organics or moderate to high levels of suspended solids will foul the membranes. Pretreatment can be used to condition the waste and reduce fouling problems. Extreme pH conditions or oxidizers in the waste solution will degrade the membrane (70).

The ability of the membrane to retain dissolved contaminants is based on molecule size, weight, and electrical charge, as well as variations of maximum pore size of the membrane; therefore retention may be difficult to predict. RO seldom exhibits "absolute" retention (77).

Evaporation produces more concentrated solutions than does RO, which is frequently used to pre-concentrate waste streams prior to evaporation.

3.21.7 Operation

Performance of the RO system typically is measured by three parameters: flux, product recovery, and rejection (78). Flux is the flowrate of permeate per unit area of membrane measured as liters per square meter (gallons per square foot) per day. The major factors influencing the sustainable flux are the physical and chemical stability of the membrane, fouling rate, and flow limits due to concentration polarization at the membrane.

Product recovery is the ratio of permeate flow to feed flow and typically is controlled by adjusting the flowrate of the reject stream leaving the RO module. Low product recoveries result in a low concentration of the metals. As product recovery increases, the metals concentra-tion of the concentrate increases, requiring an increase in pressure from the pump to overcome the osmotic pressure.

Rejection measures the degree to which the metal is prevented from passing through the membrane. Rejection increases with the ionic size and charge of the metals in the feed. Rejection is dependent on the operating pressure, conversion, and feed concentration. Typically, metals removal by RO is greater than 95 percent.

3.22 DIFFUSION DIALYSIS

3.22.1 Usefulness

Diffusion dialysis is a method to recover acids or bases for reuse by processing waste aqueous solutions that contain acids or bases contaminated with dissolved metals and organics, particulate and colloidal matter, and other dissolved or suspended non-ionic species.

3.22.2 Process Description

Diffusion dialysis is a simple ionic exchange

membrane technology that uses the concentration gradient as the driving force to achieve separation of acids or bases from waste solutions (see Figure 3-22). Anion exchange membranes allow the passage of anions only (e.g., Cl, NO_3, F, SO_4^{2-}), and cation exchange membranes allow the passage of cations only (e.g., Na^+, NH_4^+). The membranes are impermeable to nonionic species, particulates, colloids, and even organics. The only exception to the impermeability is the passage of hydrogen (H^+) and hydroxyl (OH) ions. Because of their small size and high mobility, hydrogen ions can pass through an anion exchange membrane that is impermeable to all other positively charged ions, and hydroxyl ions can pass through a cation exchange membrane. The process unit consists of a stack of anion exchange membranes (in acid recovery) or cation exchange membranes (in base recovery) with spaces between them (79).

3.22.3 Process Maturity

Large-scale diffusion dialysis units (760 to 7,600 L/day [200 to 2,000 gal/day]) for acid or base recovery are available and in commercial operation at several sites in the United States. Several vendors also supply small-scale (38 to 380 L/day [10 to 100 gal/day]) portable units. The units are modular, and increases in size or flowrate involve adding membranes to the stack.

3.22.4 Description of Applicable Wastes

The most frequent uses of diffusion dialysis have been to recover acid values from spent pickling liquor in steel plants and spent aluminum anodizing baths, to recover mineral acids from battery waste, and to recover caustic from chemical milling waste. Diffusion dialysis can recover acids or bases from solutions containing particulates, colloidal suspensions, dissolved metals, and dissolved organics. Dissolved organics are either nonionic or positively charged in acidic waste streams or negatively charged in basic waste streams, and they are rejected by the anion exchange or cation exchange membranes used in the recovery of acids and bases, respectively. The acid or base concentration in the feed waste streams can be as high as 30 percent by weight in diffusion dialysis. Waste solutions containing mixed acids (e.g., HF and nitric acid) also can be recovered.

3.22.5 Advantages

Diffusion dialysis is a low-pressure, low-temperature process that does not require the addition of treatment chemicals. Power is needed to run the pumps, a minimal requirement. Skid-mounted, modular, portable diffusion dialysis units are available, allowing convenient application on site. The modular design provides the ability to adjust capacity as needed. The process can handle acidic or base streams in small, drummed batches or as a continuous stream from a process. Operating costs are minimal.

3.22.6 Disadvantages and Limitations

Diffusion dialysis uses water in quantities equal to the volume of the waste stream. The product stream containing the acid or base values must be used on site or shipped for sale. The capital costs are high compared with conventional wastewater treatment equipment. The capital cost of a small (38 L/day [100 gal/day]) diffusion dialysis unit is $25,000. Membranes are susceptible to scaling and

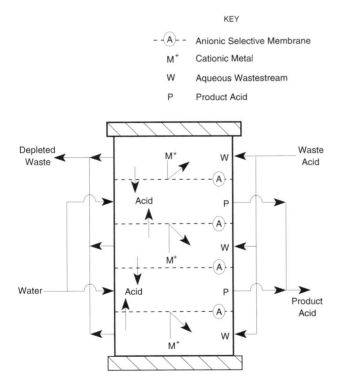

KEY

- -(A)- -   Anionic Selective Membrane
M^+ Cationic Metal
W   Aqueous Wastestream
P   Product Acid

FIGURE 3-22. Example of a diffusion dialysis cell

biological fouling. The overall process economics depend on the usable membrane life (80).

3.22.7 Operation

The aqueous waste stream and water are fed countercurrently into alternate compartments in the membrane stack. Under the influence of the concentration gradient, the acid values (consisting of H^+ ions and anions) from the waste stream pass through the anion exchange membrane and migrate to the water stream, forming the product acid. The acid-depleted waste stream containing the dissolved metals, particulates, and other nonionic species (e.g., organics) is sent for disposal. The recovered acid is either sold or used in the plant. The acid recovery rate typically is 90 to 99 percent. The flowrates of the waste stream and water stream usually are comparable. Similar flowrates and countercurrent flow assure minor or no dilution of acid concentration in the product stream relative to the feed waste stream. The unit for base recovery is similar to that for the acid system except that cation exchange membranes are used and alkali values from the waste stream migrate across the membrane to the product stream.

3.23 ELECTRODIALYSIS

3.23.1 Usefulness

Electrodialysis (ED) is a method to recover acids or bases for reuse by processing aqueous waste streams, or to concentrate and recover selected ions from aqueous waste streams containing other dissolved or suspended contaminants.

3.23.2 Process Description

In ED, ions in solution are selectively transported across ion exchange membranes under the influence of an applied direct-current field (see Figure 3-23). The ion exchange membranes are either anion selective (i.e., permeable to anions such as Cl^-, SO_4^{2-}) or cation selective (i.e., permeable to positive ions). The membrane itself acts as a barrier to the solution, to suspended particles, and to other dissolved nonionic species. An ED unit consists of hundreds of alternating anion and cation exchange membranes with spacers between them. Water and the feed containing dissolved ions and other contaminants are introduced into adjacent compartments in alternat-

ing fashion. Under the imposed polarity, the anions and cations from the feed migrate in opposite directions and concentrate in the two adjacent water-filled compartments. Because of the alternating arrangement of cation and anion membranes, the ions pass through the first membrane they encounter but are blocked by the next membrane because of their charge. The combination of migration due to the electric field and the arrangement of ion-selective membranes allows ions to concentrate in the water-filled compartments. ED units are capable of concentration factors of 10 or more. The concentrate and the diluate (feed solution depleted of most of its ions) are collected and reused (79).

3.23.3 Process Maturity

ED units are used in commercial operations to recover metal plating baths (80). Small units (38 to 190 L/day [10 to 50 gal/day]) are used in small plating job shops, and large units (380 to 3,800 L/day [100 to 1,000 gal/day]) are used in larger applications such as at aircraft maintenance depots.

3.23.4 Description of Applicable Wastes

ED is used to recover caustic from spent chemical

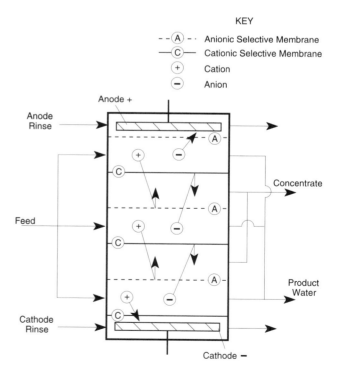

FIGURE 3-23. Example of an electrodialysis cell

milling solutions, metal values from plating rinse water, and acid from spent etchants and pickling liquors (5). ED units separate chemical values from solutions con- taining particulates and colloid suspensions. The acid or base concentrations in the feed and concentrate streams can be as high as 25 percent by weight in electrodialysis.

2.23.5 Advantages

The concentrate stream volume is small relative to the feed stream volume (by a factor of 10 or more), resulting in the recovery of chemical values in a concentrated form. Skid-mounted, modular, portable electrodialysis units are available that allow convenient application on site. The modular design provides the ability to adjust capacity as needed.

3.23.6 Disadvantages and Limitations

Electrodialysis requires a source of power (10 to 200 kW) to effect the separation. The diluate (feed depleted of its chemical values) may not meet primary treatment standards and may require additional cleanup prior to disposal. The capital costs are high compared with those for conventional wastewater treatment equipment. The membranes are susceptible to scaling and biological fouling. The overall process economics depend on the usable membrane life (80).

3.23.7 Operation

Waste stream (feed) and water are fed into alternate compartments in the ED stack. The dissolved ions (acids, bases, or metal salts) migrate from the waste stream under the influence of the electric field and concentrate in the water stream. The water stream flowrate is substantially smaller than the waste stream flowrate. The concentrate containing the recovered chemical values is collected. The ion-depleted waste stream may be reused as rinse-water or discarded, depending on the application.

3.24 EVAPORATION

3.24.1 Usefulness

Evaporation is a thermal method to concentrate and recover dissolved inorganics in aqueous materials with low concentrations of contaminants.

3.24.2 Process Description

Evaporation takes place when a liquid is heated and converts to vapor, with the liquid boiled away to leave a concentrated salt solution or slurry (see Figure 3-24). Evaporation processes may be used to recover the vaporized liquid, to form a concentrated salt solution for reuse, or to preconcentrate a salt solution for additional processing for recovery of a metal or salt (81).

3.24.3 Process Maturity

Evaporation is used commercially to reduce the volume of aqueous solutions produced by a wide variety of processes (80). Applications for reuse or recovery of valuable products are less common but have been commercialized to recover dragout from plating baths, recover contaminated acids, and produce ferric chloride from steel pickling baths (82).

3.24.4 Description of Applicable Wastes

Evaporation is best suited to processing waters containing moderate starting concentrations of dissolved salts. Waste streams containing more than about 4,000 mg/L (230 grains/gal) of metal usually can be efficiently evaporated to as high as 20 percent solids (76).

Evaporation is used to treat a variety of wastewater streams containing dissolved metal salt contaminants. For example, plating rinse wastes can be concentrated by evaporation. The evaporated water is reused for rinsing, and the concentrate is returned to the plating bath.

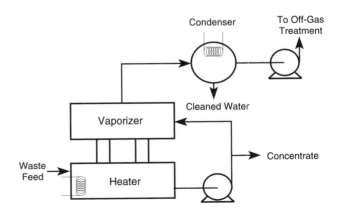

FIGURE 3-24. Example of the evaporation process

Concentration by evaporation is more difficult for wastes containing foam-stabilizing impurities such as surfactants, fine particulates, or proteins; salts that have reduced solubility at elevated temperature (e.g., calcium carbonate); compounds that decompose at elevated temperature; and high-suspended-solids concentrations.

3.24.5 Advantages

Evaporation concentrates dilute solutions of soluble salts. Volume reductions of 80 to 99 percent are possible, depending on the initial salt and suspended solids concentration and the salt solubility. The concentrated solution may be reused directly but usually is further treated by cementation, precipitation, electrowinning, or other processes to produce useable salts, metals, or brines. Evaporated water can be condensed and reused as onsite process water (5).

3.24.6 Disadvantages and Limitations

Evaporation systems generally have high capital costs and require both a significant energy input and a system to collect and treat off-gas. Suspended solids, oil, and grease cause foaming, which increases the difficulty of process operations.

3.24.7 Operation

Evaporation processes require sufficient heat energy to vaporize the liquid waste matrix. Evaporation usually is carried out either by boiling heat transfer or flash evaporation. In boiling heat transfer, steam, hot oil, or another heat transfer source is applied to transfer heat through a coil or vessel wall into the waste material. Boiling of the waste occurs at the heated wall. In flash evaporation, the waste is heated under pressure and then flows by pumping or natural circulation to a vessel at lower pressure, where boiling occurs. Common examples of evaporators using boiling heat transfer are rising film, falling film, and wiped film evaporators. Flash evaporation is accomplished in forced or natural circulation evaporators.

During the past decade, the use of atmospheric evaporators has increased for recovery of plating chemicals. The atmospheric evaporator contacts an airstream with heated solution. The air humidifies and removes water, concentrating the solution. The atmospheric evaporators have lower processing rates than conventional evaporators but use simpler, less expensive equipment and are suitable for onsite processing of small volumes of aqueous waste.

3.25 MERCURY BIOREDUCTION

3.25.1 Usefulness

Mercury bioreduction is a biochemical method to recover mercury metal for reuse by processing mercury-contaminated soils, sludges, sediments, and liquids.

3.25.2 Process Description

Bacterially mediated reduction of ionic mercury to mercury metal plays an important role in the geochemical cycle of mercury in the environment (83, 84). Two organisms, *Pseudomonas putida* (85) and *Thiobacillus ferro-oxidans* (86), have been tested for their application to the reduction and recovery of mercury wastewater. Biological activity can reduce mercury salts to metal (see Figure 3-25). Elemental mercury metal is a dense liquid with a relatively high vapor pressure, so low-energy separation methods can recover the mercury after bioreduction. In most concepts, the bioreduced mercury metal is removed from the waste matrix by air flow, then captured on activated carbon. Mercury metal is then recovered by retorting (Section 3.34). When the waste matrix viscosity is low, mercury metal may be recovered directly by gravity separation (Section 3.33).

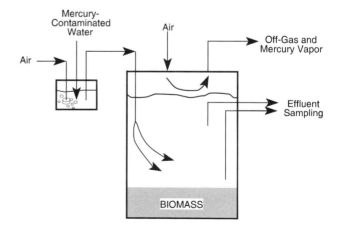

FIGURE 3-25. System to study geochemical cycling of mercury (adapted from Barkay et al. [83])

3.25.3 Process Maturity

Bioreduction of mercury salts to metal is being explored in the laboratory (87). There is no field experience with the technology.

3.25.4 Description of Applicable Wastes

The bioreduction process is potentially applicable to removal of mercury from solid or liquid wastes.

3.25.5 Advantages

Bioreduction allows mercury recovery to occur without application of heat or use of potentially hazardous chemical leaching agents.

3.25.6 Disadvantages and Limitations

Organisms can tolerate only low concentrations of mercury. Biological reactions typically proceed at slower rates than analogous chemical reactions, so longer residence times and larger reactor volumes are needed.

3.25.7 Operation

Biological detoxification of mercury-contaminated waste could be carried out in a well-mixed aerobic reactor system. The reaction is an enzyme-catalyzed reduction of ionic mercury to mercury metal. The elemental mercury could be removed from the reaction media by air stripping or gravimetric separation. If air stripping is used, mercury is captured on activated carbon, which can be treated by retorting for mercury recovery.

3.26 AMALGAMATION

3.26.1 Usefulness

Amalgamation is a chemical method to recover mercury for reuse by processing solutions of mercury salts in water.

3.26.2 Process Description

Amalgamation depends on the ability of mercury to form low-melting-point alloys with a wide variety of metals. A metal, typically zinc, that is thermodynamically able to decompose mercury compounds is contacted with a solution of mercury salt (see Figure 3-26). A chemical reaction occurs, reducing

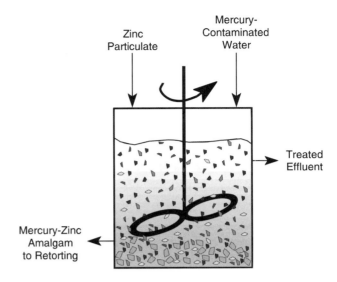

FIGURE 3-26. Example of a mercury amalgamation cell

the mercury ions to mercury metal, which then combines with the zinc to form a solid alloy. The zinc/mercury amalgam is treated by retorting to recover the mercury (88).

3.26.3 Process Maturity

In the past, mercury amalgamation was used to extract gold and silver from ores by formation of an amalgam. Mercury was retorted from the amalgam for reuse in the process, leaving gold or silver metal (89). Amalgamation currently is not applied for removal or recovery of mercury from wastewaters.

3.26.4 Description of Applicable Wastes

Amalgamation of mercury with zinc recovers mercury from salts or elemental mercury in water solution.

3.26.5 Advantages

Amalgamation recovers elemental mercury, produces an easily recoverable solid, and gives rapid reaction rates if a large surface area of sacrificial metal is available.

3.26.6 Disadvantages and Limitations

Amalgamation does not reduce the total metal content of the wastewater. The solution must be clear and free of oil, grease, and emulsified or sus-

pended matter. Noble metals in solution are precipitated by cementation and increase zinc consumption.

3.26.7 Operation

In practice, finely divided zinc is mixed with the mercury-containing solution. Good agitation is needed to ensure contact between the zinc particles and the solution. Sufficient excess zinc is required to ensure that all mercury ions are reduced and that a mercury/zinc amalgam forms. The amalgam is separated from the treated water and retorted to recover mercury metal.

3.27 CEMENTATION

3.27.1 Usefulness

Cementation is an electrochemical method to recover metals for reuse by processing from aqueous solutions.

3.27.2 Process Description

Cementation is the precipitation of a metal in solution from its salts by a displacement reaction using another, more electropositive metal (see Figure 3-27). For example, copper or silver can be displaced from solution by elemental iron.

3.27.3 Process Maturity

Cementation is a mature process used for the production of copper metal. Cementation has also been applied for the removal of copper from acid mine waters (80).

3.27.4 Description of Applicable Wastes

Cementation is applicable to recovery of dissolved noble metals from aqueous solutions. Use of aluminum to displace more-noble metals was tested and indicated the potential for removal of many salts of copper, tin and lead. The test solutions contained about 200 mg/L (12 grains/gal) of metal at a pH of 2.5. Recovery of copper complexed by ethylenediaminetetraacetic acid (EDTA) was only 51 percent. No copper was displaced from the nitrate solutions, apparently due to the acidic pH (90).

3.27.5 Advantages

Cementation recovers elemental metal, produces an easily filterable precipitate, and gives rapid reaction rates if a large surface area of sacrificial metal is available.

3.27.6 Disadvantages and Limitations

Cementation does not reduce the total metal content of the wastewater. The more-noble metal that precipitates is replaced by dissolution of the less-noble metal. Economic recovery requires an inexpensive source of less-noble metal in fine particulate form.

These disadvantages can be offset by combining cementation with electrowinning. For example, zinc can be used to electrochemically precipitate (cement) lead, copper, and cadmium from solution. The zinc is then recovered by electrowinning (Section 3.28).

3.27.7 Operation

A solution of noble metal (e.g., copper) is contacted with a less-noble metal (e.g., zinc or iron). Thorough agitation during treatment is critical for effective removal (90). The noble metal is displaced from the solution as elemental metal and is collected by filtration. Automobile shredder scrap is a common source of iron for commercial copper cementation.

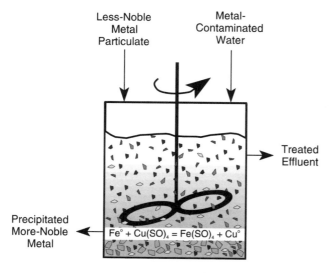

FIGURE 3-27. Example of a cementation cell

3.28 ELECTROWINNING

3.28.1 Usefulness

Electrowinning is an electrochemical method to recover elemental metal for reuse by processing moderate- to high-concentration aqueous solutions.

3.28.2 Process Description

Electrowinning uses direct current (DC) electricity applied to electrodes immersed in an aqueous solution to convert dissolved metal ions to elemental metal (see Figure 3-28). Positively charged metal ions migrate to the negative electrode, where the metal ions are reduced to elemental metal. The metal plates out on the electrode for subsequent collection and reuse.

3.28.3 Process Maturity

Electrowinning applies principles and equipment similar to those of commercial electroplating but differs in its goal: to recover metals rather than form a decorative or protective coating. In electrowinning, the metal coating appearance is unimportant, so thicker coats can be allowed to accumulate. Electrowinning is applied commercially in the mineral refining industry and for recovery of metals from spent electroplating baths.

3.28.4 Description of Applicable Wastes

Electrowinning is most effective for recovery of more-noble metals. Metals with a high electrode potential are easily reduced and deposited on the cathode. Gold and silver are ideal candidates, but cadmium, chromium, copper, lead, nickel, tin, and zinc also can be recovered using a higher voltage (80, 91). Electrowinning will remove metals from solutions containing chelating agents, which are difficult to recover by physical or chemical processes (70).

3.28.5 Advantages

Electrowinning recovers metals from aqueous solutions without requiring further processing and without generating any metal-containing sludge or process residuals. The residual metal in the liquid effluent from a well-designed and well-operated electrowinning cell is significantly reduced, but ion exchange polishing of the effluent may be required (70, 92).

3.28.6 Disadvantages and Limitations

Electrowinning is most efficient when applied to concentrated solutions. For example, ion exchange resins can be used to remove metals, which are then concentrated in the solution used to regenerate the resin (Section 3.19). The regeneration solution is treated by electrowinning to recover the metal.

Electrowinning typically does not remove metals in solution to acceptable limits for discharge. Ion exchange often is used to further reduce the metal concentration in a solution after electrowinning. The regeneration chemicals from the ion exchange process also are treated in the electrowinning cell to complete metal recovery (93). Cementation (Section 3.27) also is used to polish water treated by electrowinning (91).

Electrowinning is best applied to solutions with one metal contaminant. Selecting the electrode potential can control the type of metal deposited on the cathode. For example, more-noble metals are removed at lower applied voltage. Noble metals can be recovered from mixed-metal baths, but base metals

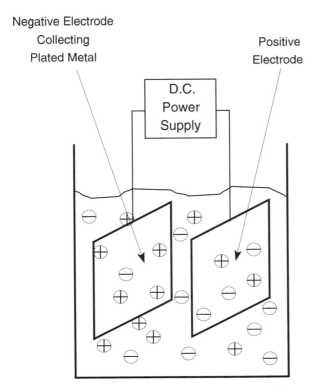

Negative Electrode
Collecting
Plated Metal

Positive
Electrode

D.C.
Power
Supply

FIGURE 3-28. Example of an electrowinning cell

are more difficult to separate. Electrowinning from mixed-metal systems increases the complexity of operation.

Adverse reactions can occur in an electrochemical cell depending on the impurities present in the waste (80). An acid mist is generated over the electrowinning cell. The acid off-gas must be collected and treated (70).

3.28.7 Operation

An electrowinning metal recovery system requires an electroplating tank, a direct current power source, electrodes, and support equipment. Solution flow should be maintained past the cathode by either rotating the cathode or moving the solution by pumping or air agitation (94). The cathode is often a thin sheet of the metal being recovered and may be reticulated to increase electrode surface area. High electrode surface area decreases the local current density, increases current efficiency, and reduces electrode corrosion. When a cathode is loaded with recovered metal, it is removed and replaced. The metal-coated cathode is sold for the metal value. Flat titanium cathodes also are used. With titanium cathodes, the electrowon metal is stripped off in thin sheets. Less expensive cathode designs such as metal-coated plastic sheet or mesh may be used, but the value of the recovered metal is reduced by embedded plastic. The anode is made of a conductive and corrosion-resistant material such as stainless steel or carbon (5).

Electrowinning from dilute solutions is less efficient due to low ion diffusion rates. The efficiency of electrowinning metals from dilute solutions can be improved by heating the solution, by agitation, and by using a large cathode area. Salt may need to be added to maintain minimal conductivity in the solution being treated (82).

3.29 CHEMICAL LEACHING

3.29.1 Usefulness

Leaching is a chemical method to recover metals or metal compounds for reuse by processing solids and sludges containing low to moderate concentrations of contaminants. A description of desirable properties for feed materials to chemical leaching is given in Section 4.8.

3.29.2 Process Description

Chemical leaching transfers metals from a solid matrix into the leaching solution (see Figure 3-29). Solution processing methods are then used to regenerate the leachant and recover a useful metal or salt. The combination of chemical leaching and leachant regeneration is known as hydrometallurgical processing. Hydrometallurgical processing typically includes one or more of the following four steps:

- Dissolution of the desired metal
- Purification and/or concentration of the metal
- Recovery of the metal or a metal salt
- Regeneration of the leaching solution

3.29.3 Process Maturity

Chemical leaching is developed at the commercial scale for recovery of metals from various sludges, catalysts, and other solid matrices. Several RCRA-permitted facilities are available for processing leachable metal characteristic wastes and listed wastewater treatment sludges (95). Pilot-scale tests have demonstrated lead recovery from contaminated soil at Superfund sites by acid extraction followed by regeneration of the extraction solution (96-98).

3.29.4 Description of Applicable Wastes

Chemical leaching and hydrometallurgical processing can be applied to a variety of solid and sludge wastes. Wastes containing a high concentration of one metal in one valence state are preferred (99). Waste streams processed include wastewater treatment sludges (e.g., plating operations [F006], metal finishing, and electronic circuit board etching), baghouse dust, and spent catalyst. The metals reclaimed include chromium, nickel, copper, zinc, lead, cadmium, tin, cobalt, vanadium, titanium, molybdenum, gold, silver, palladium, and platinum.

Chemical leaching for mercury recovery is a new and growing technology area. Several chemical leaching formulations have been developed to remove mercury from contaminated soils (87). Processes to recover lead by acid leaching followed by electrowinning are being developed (100, 101).

Products are often metal salts. For example, hydroxide plating or etching sludge can be converted to metal salts such as copper chloride, copper ammonium chloride, or nickel carbonate (102). For cat-

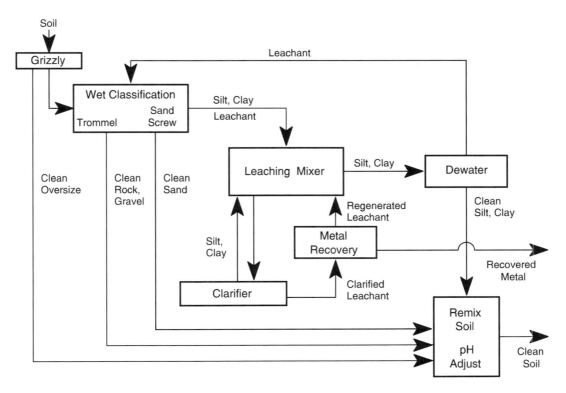

FIGURE 3-29. Example of the chemical leaching process (adapted from U.S. EPA [24])

alysts, metals and substrate materials can be converted by leaching and solution processing into products such as nickel-copper-cobalt concentrate, alumina trihydrate, chromium oxide, molybdenum trioxide, and vanadium pentoxide (103).

3.29.5 Advantages

Hydrometallurgical processes recover metal contaminants to produce metal or metal salts directly or upgrade low-concentration materials to allow metal recovery in secondary smelters. For some wastes, the process also may recover the matrix in a useful form, leaving no residue requiring disposal. Hydrometallurgy usually is more efficient than pyrometallurgy if the metal concentration is low (from the percent range down to parts per million). Hydrometallurgical processing can require less-capital-intensive facilities than pyrometallurgical processing, but economies of scale still apply.

3.29.6 Disadvantages and Limitations

The chemical leaching operation produces large volumes of leach solution. These solutions typically are regenerated for reuse to leach the next batch of material. Reuse is required both to recover the economic value of the leaching chemicals and to avoid the environmental impacts associated with treatment and discharge of waste solutions.

Leaching of soils with a high clay content leaves fine, suspended particulate in the leachant. This fine particulate complicates further processing of the leachant and is difficult to remove.

3.29.7 Operation

Hydrometallurgy uses aqueous and/or organic solvents to dissolve a metal from the solid matrix. (For more information on liquid ion exchange/solvent extraction of metals, see Section 3.20.) The dissolution process is called leaching. The leaching solution is chosen, based on the types of metals and compounds present in the matrix being treated, to maximize recovery of valuable metals while minimizing dissolution of unwanted species in the matrix. The leaching chemistry often relies on formation of metal complexes. However, a newly developing branch of hydrometallurgy, called supercritical extraction, relies on the unusual solvent power of water, CO_2, and organics above the critical pressure and temperature to affect the selective dissolution.

Once the metal is in solution, further processing typically is required to remove impurities, increase the metal concentration, or both. The full range of classical solution processing methods is available for upgrading the leach solution. The most commonly used methods are precipitation (Section 3.18), liquid ion exchange (Section 3.20), and conventional resin ion exchange (Section 3.19).

The concentrated and purified metal-containing solution typically requires further treatment to produce a marketable product. In some cases, the metal salt or complex is reduced to native metal. Reduction to metal can be accomplished by electrowinning or by reducing the metal using a reducing gas such as hydrogen. Alternatively, the end product of the hydrometallurgical process may be a metal salt. Chemical processing converts the compound in solution to a more marketable oxide or salt form.

3.30 VITRIFICATION

3.30.1 Usefulness

Conversion to a ceramic product is a thermal method to form useful products from slags and sludges with low concentrations of metal contaminants alone or mixed with organics. Ceramic products can range from high-value materials such as abrasives or architectural stone to low-value materials such as aggregate. Some characteristics of high- and low-value ceramic products are outlined in Sections 4.9 and 4.12, respectively.

3.30.2 Process Description

This technology uses heating to promote oxidation, sintering, and melting, thus transforming a broad spectrum of wastes into a glasslike or rocklike material. The melting energy can be derived from the oxidation of materials in the feed supplemented by combustion of fossil fuels or electrical heating. The process typically collects particulate in the off-gas system and returns the particulate to the melter feed so secondary waste generation is minimized. The discharged solid can be formed into ceramic products (see Figure 3-30).

3.30.3 Process Maturity

Vitrification/sintering to form a stable glasslike or rocklike solid is a commercially available technology (105).

3.30.4 Description of Applicable Wastes

Waste materials amenable to treatment include filter cakes, foundry sand, ash, and sludge. The process treats inorganic wastes containing cadmium, chromium, cobalt, copper, lead, nickel, vanadium, or zinc. Examples of suitable wastes include sludge from wastewater treatment, electric arc furnace off-gas treatment residues, and baghouse dust (104).

The presence of volatile metals in the waste complicates vitrification processing due to production of metal vapors in the off-gas. Metals such as mercury or beryllium that volatilize under process conditions may not be amenable to treatment. Wastes containing arsenic require some combination of pretreatment and special processing conditions and off-gas treatment systems to minimize arsenic volatilization.

Selected waste streams can be converted into high-value ceramic products such as abrasives, decorative architectural stone, or refractory. Wastes suitable for these processes typically are hydroxide sludge from treatment of plating or etching baths containing a single metal contaminant. The process has been applied commercially to F006 wastes (106).

Processes have been demonstrated for thermal conversion of a variety of silicate matrices containing metal and organic contaminants into intermediate-value ceramic products. For example, fly ash (107), incinerator ash (108), iron foundry slag (109), and petroleum-contaminated soils (110, 111) have been used for the manufacture of bricks. In addition, spent abrasive blasting media have been used to replace sand in the manufacture of bricks and mortar.

Low-value construction aggregate and stone can be produced from a variety of waste materials. Examples include concrete aggregate produced from fly ash (112, 113) or fill material produced by vitrification of metal-containing wastes.

3.30.5 Advantages

The discharged product is a chemically durable material that typically passes the Toxicity Characteristic Leaching Procedure (TCLP) test as nonhazardous. The process provides volume reduction (40 percent for soils to greater than 99 percent for com-

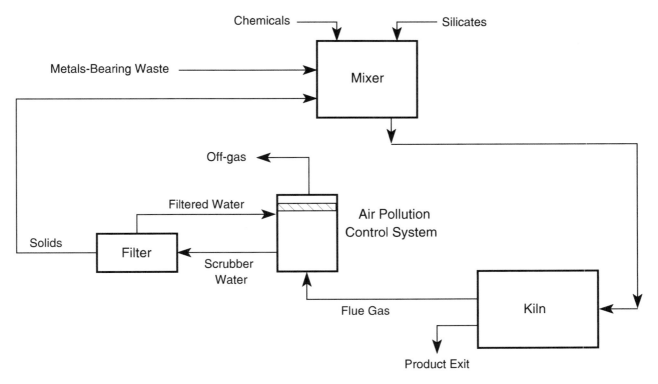

FIGURE 3-30. Example of the vitrification procss (adapted from U.S. Air Force [103]).

bustibles). The high operating temperature destroys organic contaminants in the waste (70).

3.30.6 Disadvantages and Limitations

The thermal conversion process is capital and energy intensive. Revenue is unlikely to equal processing costs, even for waste streams that form a high-value product. The main economic advantage is avoided disposal costs.

Thermal processing generates large volumes of off-gas that must be controlled and cleaned. Volatile metals in the waste, particularly arsenic, beryllium, or mercury, complicate processing.

3.30.7 Operation

Ceramic products may be formed by either sintering or melting. In both processes, prepared waste material is heated to form the ceramic. Most thermal treatment processes require feed material to be within a narrow particle size range. Size reduction, pelletization, or both processes usually are needed to obtain the required size.

In sintering, the waste is prepared by mixing with clay or other silicate and possibly water and additives. The resulting mix is pressed or extruded to form pellets, which are treated at high temperature but below the bulk melting temperature. Particles join to form a solid ceramic piece.

Vitrification processes also require feed preparation. Chemical additions and mixing may be used to promote oxidation-reduction reactions that improve the properties and stability of the final product. Silica sources such as sand or clay also may be needed. Vitrification processes operate by heating the pretreated waste to melting temperatures. The molten, treated waste flow exits from the melter into a waste forming or quenching step. The melt can be formed in a sand-coated mold or quenched in a water bath, depending on the type of product required.

Gases released from the thermal treatment unit are processed through an emission control system. Particulates may form due to carryover, metal fuming, or anion fuming. The particulates are removed by knockout boxes, scrubbers, and/or venturi separators. Particulates are separated from the scrubbing fluid by filtration and are returned to the treatment system. Acid gases, such as sulfur dioxide from sulfates, are removed by scrubbing with sodium hydroxide.

3.31 PYROMETALLURGICAL METAL RECOVERY

3.31.1 Usefulness

Pyrometallurgical processing is a thermal method to recover metals or metal compounds for reuse by processing solids and sludges with percent concentrations of metal contaminants. A description of desirable properties for feed material to a pyrometallurgical metal recovery process is given in Section 4.7. A case study of processing Superfund site wastes in a secondary lead smelter is described in Section 5.4.

3.31.2 Process Description

Pyrometallurgy is a broad term covering techniques for processing metals at elevated temperature. Processing at elevated temperature increases the rate of reaction and reduces the reactor volume per unit output (see Figure 3-31). Elevated temperature also may benefit the reaction equilibrium. Pyrometallurgy offers a well-developed and powerful collection of tools for recovery of metals from waste materials. Three general types of pyrometallurgical processes are in use:

- Pretreatment of material as preparation for further processing.

- Treatment of material to convert metal compounds to elemental metal or matte and to reject undesirable components.

- Subsequent treatment to upgrade the metals or matte.

These operations may be used singly, in sequence, or in combination with physical, hydrometallurgical, biological, or electrometallurgical processing depending on the types of material processed. The three types of pyrometallurgical processes generally use different equipment and approaches.

3.31.3 Process Maturity

Pyrometallurgical processing is developed at the commercial scale for recovery of cadmium, lead, and zinc from K061 (RCRA waste code for Electric Arc Furnace emission control dust/sludge) and a variety of metal-containing silicate and sludge wastes.

3.31.4 Description of Applicable Wastes

Pyrometallurgical processing typically is used to process large volumes of solid or low-moisture sludge containing percent levels of metals. Processing capacity is available for a variety of metal leachability characteristic wastes, F006-listed waste, and other metal-containing soils, slags, dusts, sludges, ashes, or catalysts. The types of materials processed include wastes from electroplating, electropolishing, metal finishing, brass and steel foundries, galvanizing, zinc diecasting, nickel-cadmium and iron-nickel batteries, chromium-magnesite refractories, waste magnesium powders and machinings, and pot liner from aluminum smelters (114-119).

Some facilities may accept wastes containing trace quantities of metals if the matrix is a good source of silica or alkali for flux or a carbon source for metal reduction. Processing ash from incineration of municipal wastewater treatment sludge provides silica as a flux and allows recovery of trace quantities of gold and silver. Pyrometallurgical processors also may accept used foundry sand for metals and silica, lime residues from boiler cleaning or dolomitic refractories for metals and alkali, or carbon brick and coke fines for carbon.

3.31.5 Advantages

Pyrometallurgy is an extractive treatment approach that can recover metals or metal salts for reuse. The high operating temperature destroys organic contaminants in the waste (70). The volume of slag residual resulting from the process typically is smaller than the initial waste volume. In most cases, the hazardous metal concentration in the slag is low and the slag is leach resistant, so it may be reused as a low-value aggregate product.

3.31.6 Disadvantages and Limitations

Pyrometallurgical processing is applicable only to specific types of wastes. Success depends both on the types and concentrations of metals present and the physical and chemical form of the matrix.

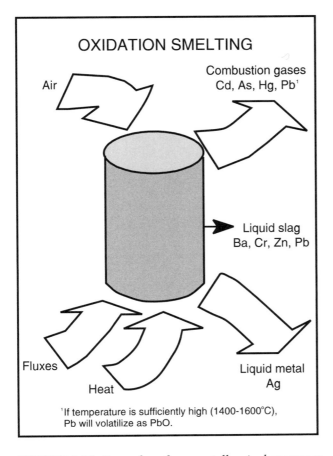

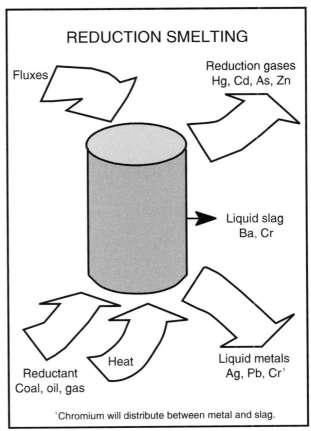

FIGURE 3-31. Examples of pyrometallurgical processes

The process is capital and energy intensive. Processing can be profitable or break even when the waste contains a high concentration of valuable metal or can be processed as a small addition to an existing feedstream. For other waste types, pyrometallurgical processing is costly.

Pyrometallurgical processing generates large volumes of off-gas that must be controlled and cleaned. Volatile metals in the waste, particularly arsenic, complicate processing.

3.31.7 Operation

Oxidation is often used as a pyrometallurgical pretreatment to convert sulfide materials to oxides. Normally oxidation is carried out as a gas-solid phase contact of air passing over fine particulate material in a multihearth or fluidized bed reactor. Pretreatment also is used to selectively metallize the feed for subsequent leaching.

Pyrometallurgical processing to convert metal compounds to metal usually requires a reducing agent, fluxing agents to facilitate melting and slag off impurities, and a heat source. The fluid mass often is called a melt, although the operating temperature, while quite high, often is below the melting points of the refractory compounds being processed. The fluid forms as a low-melting-point material due to the chemistry of the melt. An acceptable melting point is achieved by addition of fluxing agents, such as calcium oxide, or by appropriate blending of feedstocks.

Carbon is the most commonly used reducing agent for base metal compounds. The carbon typically reacts to produce carbon monoxide and carbon dioxide while forming free metal or matte. Combustion of the carbon also can provide the required heat input. Feedstocks, such as sulfides or organic-laden wastes, react exothermically, thus providing some heat input as well.

The most common fluxing agents in mineral smelting are silica and limestone. The flux is added along with the reducing agent to produce a molten mass. A wide variety of molten salts, molten metals, or other fluxing systems are used for special processing situations.

Separation of the metal from the undesirable waste components typically is accomplished by physical action based on phase separations. As the metal salts react with the reducing agent to form metal or matte, the nonmetallic portions of the ore combine with the flux to form a slag. Volatile metals such as zinc or cadmium vaporize and are collected by condensation or oxidation from the off-gas, usually as oxides due to combustion of metal fume in the flue. Dense, nonvolatile metals can be separated from the less dense silicate slag by gravity draining of metallics from the bottom of the reaction vessel. Slag oxides are tapped from a more elevated taphole.

Pyrometallurgical processes for final purification (refining) usually take the form of selective volatilization, drossing, or liquation. In some cases, ores consist of less-volatile metals such as lead and iron mixed with volatile metals such as zinc, cadmium, or arsenic. For other metal systems, chemical reactions are needed to form volatile species to allow separation of product metals and impurities.

3.32 CEMENT RAW MATERIALS

3.32.1 Usefulness

Use of inorganic wastes as raw material in cement manufacture is a thermal method to recover inorganics, mainly aluminum, iron, or silica, from solid materials with low concentrations of contaminants. Burning of mainly organic-containing materials in a cement kiln for heating value is discussed in Section 3.3. A description of desirable properties for cement raw materials is given in Section 4.10. A case study using spent sand blasting media for cement raw materials is described in Section 5.2.

3.32.2 Process Description

In this process, waste materials are fed to a cement kiln as a substitute for raw materials such as limestone, shale, clay, or sand. The primary constituents of cement are silica, calcium, aluminum, and iron. Inside the cement kiln, the raw material substitutes undergo chemical and physical reactions at temperatures that progressively reach 1,480°C (2,700°F) to form cement clinker (see Figure 3-32). Inorganic contaminants are bound into the lattice structure of the cement crystals.

3.32.3 Process Maturity

Nonhazardous silicate and aluminate wastes are used as raw material substitutes in Portland cement manufacture on a commercial scale. Application to wastes containing RCRA metals may be possible, but commercial application is limited by the requirements of the Boiler and Industrial Furnace regulations.

3.32.4 Description of Applicable Wastes

The primary raw materials of interest are silica, calcium, aluminum, and iron. Good candidates for raw materials substitution typically contain 95 percent or more of these constituents. Examples of acceptable feed materials include the following sources:

Alumina sources:
- Catalysts
- Ceramics and refractories
- Coal ash
- Adsorbents for gases and vapors
- Aluminum potliner waste

Calcium sources:
- Lime sludges

Iron sources:
- Foundry baghouse residuals
- Iron mill scale

Silica sources:
- Abrasives
- Ceramics
- Clay filters and sludges
- Foundry sand
- Sand blast media
- Water filtration media

3.32.5 Advantages

Cement kilns provide high operating temperatures and long residence times to maximize the im-

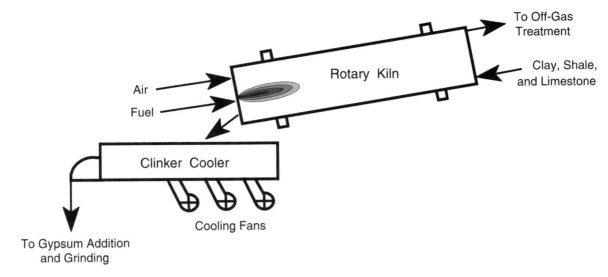

FIGURE 3-32. Example of cement kiln operation

mobilization of metal contaminants into the cement mineral structure. The high-alkali reserve of the cement clinker reacts to form alkali chlorides (sodium, potassium, calcium) that prevent the evolution of acidic vapors in the off-gas. The chloride content must be limited, however, to avoid vapor production and to prevent soluble chlorides from degrading the setting rate of the cement product (119).

3.32.6 Disadvantages and Limitations

Both combustion to heat the raw materials and decomposition reactions during formation of cement clinker generate large volumes of off-gas that must be controlled and cleaned.

3.32.7 Operation

Raw material burning typically is done in a rotary kiln. The kiln rotates around an inclined axis. The raw materials enter the raised end of the kiln and travel down the incline to the lower end. The kiln is heated by combustion of coal, gas, or oil in the kiln. As the raw materials move through the inclined, rotating kiln, they heat to temperatures in the area of 1,480°C (2,700°F). The high temperature causes physical and chemical changes, such as (8):

- Partial fusion of the feed materials

- Evaporation of free water

- Evolution of carbon dioxide from carbonates

During burning, lime combines with silica, alumina, and iron to form the desired cement com-

pounds. The heating results in the cement clinker. Clinker consists of granular solids with sizes ranging from fine sand to walnut size. The clinker is rapidly cooled, mixed with additives such as gypsum, and ground to a fine powder to produce the final cement product.

Portland cement product is produced by heating mixtures containing lime, silica, alumina, and iron oxide to form clinker, which is then ground. About 3 to 5 percent of calcium sulfate, usually as gypsum or anhydrite, is added during grinding of the clinker. The gypsum aids the grinding process and helps control the curing rate of the cement product (120). The gypsum is intermixed during grinding of the clinker. The main constituents of Portland cement typically are tricalcium silicate, dicalcium silicate, tricalcium aluminate, and tetracalcium aluminoferrite.

3.33 PHYSICAL SEPARATION

3.33.1 Usefulness

Physical separation uses physical differences to concentrate and recover solids suspended in water or mixed with other solids. A case study of lead recovery from soil at a small-arms practice range using physical separation is described in Section 5.3.

3.33.2 Process Description

Physical separation/concentration involves separating different types of particles based on physical characteristics. Most physical separation operations are based on one of four characteristics:

- Particle size (filtration or microfiltration)
- Particle density (sedimentation or centrifugation)
- Magnetic properties (magnetic separation)
- Surface properties (flotation)

3.33.3 Process Maturity

Application of physical separation methods is well established in the ore-processing industry. Physical separation provides a low-cost means of rejecting undesirable rock and debris, thus increasing the concentration of metal and reducing the volume to process (see Figure 3-33). Mining experience is now being extended to full-scale application of physical separation at Superfund sites.

3.33.4 Description of Applicable Wastes

Physical separation is applicable to recovery of metals from soils, sediments, or slags in either of two situations.

First, discrete metal particles in soil can be recovered based on size, density, or other properties. For example, mercury metal can be recovered by gravity separation, lead fragments can be separated by screening or by gravity methods, and high-value metals (e.g., gold or silver) can be recovered by membrane filtration. The most common applications are size and gravity recovery of lead in firing range or battery breaking site soils and gravity recovery of

elemental mercury from contaminated soils (87, 97, 121).

Second, metals present in elemental or salt form may be sorbed or otherwise associated with a particular size fraction of soil material. Materials tend to sorb onto the fine clay and silt in soil. Physical separation can divide sand and gravel from clay and silt, yielding a smaller volume of material with a higher contaminant concentration. The upgraded material can then be processed by techniques such as pyrometallurgy or chemical leaching to recover products.

3.33.5 Advantages

Physical separation allows recovery of metals or reduction of the volume to be treated using simple, low-cost equipment that is easily available from a wide variety of vendors.

3.33.6 Disadvantages and Limitations

Physical separation requires that the desired component be present in higher concentration in a phase having different physical properties than the bulk material. Separation methods applied to dry the material (e.g., screening) generate dust.

3.33.7 Operation

The general characteristics of some common particle separation techniques are summarized in Table 3-2.

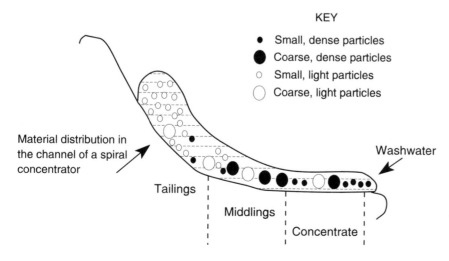

KEY

- ● Small, dense particles
- ⬤ Coarse, dense particles
- ○ Small, light particles
- ◯ Coarse, light particles

Material distribution in the channel of a spiral concentrator

Washwater

Tailings

Middlings

Concentrate

FIGURE 3-33.
Cross section showing
particle distribution
in a spiral concentrator
channel.

TABLE 3-2
Particle Separation Techniques (122, 123)

	Technique				
	Screen	**Classification by Settling Velocity**	**Gravity Separation**	**Magnetic Separation**	**Flotation**
Basic Principle	Various diameter openings allowing passage of particles with different effective sizes	Different settling rates due to particle density size, or shape	Separation due to density differences	Magnetic susceptibility	Particle attraction to bubbles due to their surface properties
Major Advantage	High-throughput continuous processing with simple inexpensive equipment	High-throughput continuous processing with simple inexpensive equipment	High-throughput continuous processing with simple inexpensive equipment	Recovery of a wide variety of materials when high gradient fields are used	Effectiveness for fine particles
Limitations	Screens can plug; fine screens are fragile; dry screening produces dust	Process is difficult when high proportions of clay, silt, and humic materials are present	Process is difficult when high proportions of clay, silt, and humic materials are present	Process entails high capital and operating costs	Particulate must be present at low concentrations
Typical Implementation	Screens, sieves, or trommels (wet or dry)	Clarifier, elutriator, hydrocyclone	Shaking table, spiral concentrator, jig	Electromagnets, magnetic filters	Air flotation columns or cells

3.34 MERCURY ROASTING AND RETORTING

3.34.1 Usefulness

Mercury roasting and retorting is a thermal method to recover mercury metal for reuse by processing mercury-containing solids and sludges.

3.34.2 Process Description

Relatively few metal oxides convert easily to the metallic state in the presence of oxygen. Reduction to metal typically requires the presence of a reducing agent such as carbon, as well as elevated temperatures; mercury is one of the few exceptions. Many mercury compounds convert to metal at an atmospheric pressure and temperature of 300°C (570°F) or less. Mercury also is substantially more volatile than most metals, with a boiling point of 357°C (675°F). As a result, mercury and inorganic mercury compounds can be separated from solids by roasting and retorting more easily than most metals can (87).

Retorting is a decomposition and volatilization process to form and then volatilize elemental mercury (see Figure 3-34). Waste is heated in a vacuum chamber, usually indirectly, in the absence of air to a temperature above the boiling point of mercury. Heating normally is done as a batch process. The exhausted mercury is collected by condensation, water scrubbing, or carbon adsorption. Mercury roasting is the process of heating mercury-containing materials in air, typically to convert sulfides to oxides in preparation for retorting.

3.34.3 Process Maturity

A commercial infrastructure has been established for recycling mercury-containing scrap and waste materials (125). Recovery of mercury from soils by thermal treatment has been practiced on a commercial scale (87, 126). Industrial production of mercury from recycling of secondary sources amounted to 176 metric tons (194 tons) in 1990 (127).

3.34.4 Description of Applicable Wastes

Dirt, soils, and sludge-like material can be

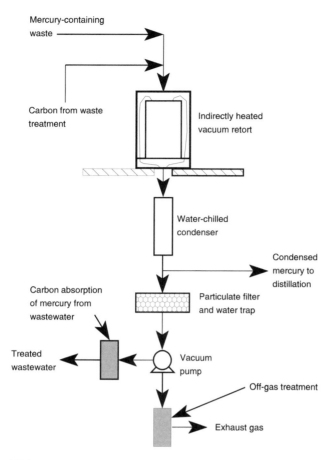

Mercury-containing waste

Carbon from waste treatment

Indirectly heated vacuum retort

Water-chilled condenser

Condensed mercury to distillation

Carbon absorption of mercury from wastewater

Particulate filter and water trap

Treated wastewater

Vacuum pump

Off-gas treatment

Exhaust gas

FIGURE 3-34. Example of the mercury retorting process (adapted from Lawrence [123])

processed if the water content is below about 40 percent. If the mercury is in solution, the mercury must be collected as a solid by precipitation or adsorption onto activated carbon. As with the sludge feed, the collected solids must contain less than about 40 percent water. Land Disposal Restriction treatment standards for various RCRA nonwastewater mercury-containing wastes with concentrations of more than 260 mg/kg (260 ppm) mercury were developed based on thermal treatment to recover mercury as the Best Demonstrated Treatment Technology.

3.34.5 Advantages

Roasting and retorting each provide an effective means to recover mercury from a variety of mercury-containing solids. Both fixed facilities and transportable units are available (128).

3.34.6 Disadvantages and Limitations

Roasting and retorting generate an off-gas stream that must be controlled and cleaned. Volatile metals in the waste, particularly arsenic, complicate processing.

3.34.7 Operation

In the retorting process, solids contaminated with mercury are placed in a vacuum-tight chamber. Following closure of the chamber, a vacuum is established and heat is applied. Operating the retort under vacuum helps collect and control mercury emissions from the process. The materials in the chamber are subjected to temperatures in excess of 700°C (1,300°F). Mercury is vaporized from the material, withdrawn, and collected. The mercury can be further purified by distillation. The mercury-free solids are transported to other facilities for recovery of other metals, if possible.

Typical feed materials include metal and glass materials. Most plastics can be processed, but PVC and other halogen-containing materials must be minimized due to the potential for generating corrosive or volatile materials during heating in the retort. Volatile or reactive metals such as lithium, arsenic, and thallium also are not allowed in the process (124).

3.35 MERCURY DISTILLATION

3.35.1 Usefulness

Mercury distillation is a thermal method to recover high-purity mercury metal for reuse by processing slightly contaminated liquid elemental mercury.

3.35.2 Process Description

Mercury distillation relies on mercury's relatively low boiling point to allow purification of slightly contaminated liquid mercury to very high levels of purity. Mercury and volatile impurities boil off in a vacuum retort, leaving nonvolatile impurities in the retort. The mercury vapors are then distilled to concentrate volatile impurities in a mercury heel to produce high-purity mercury (see Figure 3-35). Multiple passes through retorting and distillation can produce 99.9999 percent or higher purity mercury (125).

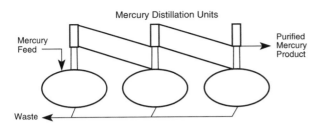

Mercury Distillation Units

Mercury Feed →

Purified Mercury Product →

Waste ←

FIGURE 3-35. Example of the mercury distillation process (adapted from Lawrence [123])

3.35.3 Process Maturity

Distillation is commercially available for final cleanup of slightly contaminated liquid mercury metal.

3.35.4 Description of Applicable Wastes

Feed to a mercury distillation process typically is free-flowing mercury liquid showing a shiny surface and no visible water, glass fragments, or other solids. Physical separation may be used to remove solids to allow distillation of the mercury. Wastes containing lead, cadmium, or arsenic usually are unacceptable for distillation.

Typical waste sources are mercury metal liberated from solid by retorting, electronic scrap, and impure, used mercury.

3.35.5 Advantages

Distillation can be applied to clean mercury recovered by roasting or retorting (Section 3.34).

3.35.6 Disadvantages and Limitations

The process generates an off-gas stream that must be controlled and cleaned. Volatile metals in the waste, particularly arsenic, complicate processing.

3.35.7 Operation

Mercury distillation typically is carried out in small batch vacuum systems. Vacuum distillation reduces the required operating temperatures and helps collect and control mercury emissions from the process. The usual processing capacity for a batch is 208 L (55 gal). Higher purity mercury is produced by redistillation and/or washing. Specially designed distillation operations can produce 99.99999 percent

("7 nines") pure mercury with quadruple distillation. The more common approach is triple distillation followed by washing the mercury with dilute nitric acid to yield 99.9999 percent ("6 nines") pure mercury.

3.36 DECONTAMINATION AND DISASSEMBLY

3.36.1 Usefulness

Decontamination and disassembly uses mechanical and chemical methods to clean and disassemble process equipment and structures to allow recovery of metals or inorganic materials for reuse. A source for specifications describing bulk metals for recycling is given in Section 4.6.

3.36.2 Process Description

The process includes the dismantling of equipment; decontamination of walls, ceilings, and floors; and the cleaning or removal of utilities such as drains, ductwork, filters, vents, and electrical conduits. The building then may be either reused or dismantled. The major steps in decontamination and disassembly include:

- Preparation of equipment for removal.
- Disassembly of equipment and fixtures.
- Equipment decontamination.
- Metal cutting, including electrical conduit, drains, and vent systems.
- Metal decontamination.
- Floor tile removal.
- Decontamination of floors, walls, and ceilings.
- Concrete floor cutting and excavation of soil to remove subfloor drains (if necessary).

3.36.3 Process Maturity

A multitude of cleaning and disassembly methods are commercially available for equipment, metals, and other materials such as glass, brick, wood, rubber, and concrete. Cleaning technologies range from detergents and water, grinding methods, and chemical treatments to the use of lasers, CO_2 pellet blasting (see Figure 3-36), and other methods that produce a minimum of additional waste. Disassembly methods include wire, flame, and water cutters;

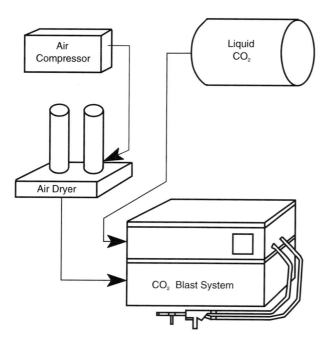

FIGURE 3-36. Example of decontamination apparatus

saws; shears; nibblers; and large impact equipment. Disposal options for waste can be costly; therefore, waste minimization increases in importance. Cleaning methods should be selected to avoid production of or to minimize the volume of hazardous waste. An understanding of the Land Disposal Restrictions can assist facility management in the selection of the most effective treatment and/or disposal methods (129).

3.36.4 Description of Applicable Wastes

Different methods are effective on different contaminated materials. Typical methods for cleaning metals are high-pressure water blasting, solvents, and other media blasting (plastic, wheat starch, sodium bicarbonate, carbon dioxide pellets, ice pellets), cryogenic, and thermal treatments. Advanced paint removal technologies were reviewed in a recent EPA report (130). The CO_2 blasting method was used successfully at a Superfund site to clean building and tank surfaces contaminated with mercury and heavy metals (131). Concrete can be cleaned with these methods and others, such as by grinding and milling, or using strippable coatings and foam cleaners. The strippable coatings, foams, and blasting methods are effective on painted metal

and concrete surfaces, but electrical conduit is most effectively cleaned with a dry method. Drains and complex, hard-to-reach surfaces can be cleaned with solvents and foams.

Concrete cutting methods include core stitch drilling, diamond wire cutting, flame and water cutting, sawing, and use of impact equipment, such as a backhoe-mounted ram or paving breaker (hammer-like devices). Metal and equipment can be decommissioned using arc saws, shears, nibblers, torches, water cutting equipment, power saws, band saws, or guillotine saws.

3.36.5 Advantages

Surface decontamination methods, such as carbon dioxide pellet blasting and laser heating, add a minimum amount of waste to the pretreated quantity. Abrasive blasting and thermal cleaning methods provide high throughput rates. For cutting structures, wire cutting is a faster, more precise cutting method than traditional cutting methods. Saws provide versatility by using specific blades for various materials.

3.36.6 Disadvantages and Limitations

Abrasive blasting methods can damage the underlying surface. Most cleaning techniques require varying levels of training and respiratory, eye, skin, and ear protection. Water blasting, plastic media blasting, solvents, foams, and detergents all add to the pretreated quantity of waste. Most cutting techniques also require dust and contamination control before, during, and/or after cutting operations. Some equipment, such as large saws, wire cutters, shears, and coring machines, can be heavy and cumbersome to use.

Decontamination and disassembly processing can reveal asbestos in many forms. Asbestos-containing material was used as insulation, in siding and shingles, and in laboratory hood and equipment linings (transite). Site surveys should include identification and characterization of asbestos-containing materials and appropriate planning for decontamination and disassembly operations.

3.36.7 Operation

The types of equipment and materials and the time required for decontamination vary depending on which method is used. Solvents may be used in a self-contained unit or may be circulated through pipes and drains. Cleaning times range from a few minutes to a few days. Blasting equipment generally consists of a pumping mechanism, vacuum mechanism, and possibly a treatment system for recycling the blasting media. Cleaning times range from 0.5 to 35 m²/hr (5 to 350 ft²/hr).

Cutting and grinding techniques may require pretreatment or a tent for dust control, as well as stabilization of surrounding material. A vacuum system also may be used for cleanup and containment of the waste. The time required for operation depends on the material and the equipment chosen, and can range from 0.1 to 1.8 m²/hr (1 to 20 ft²/hr).

3.37 RECYCLING TRANSFORMERS AND BALLASTS

3.37.1 Usefulness

Electrical equipment containing PCB-dielectric oils can be processed to destroy the PCBs, allowing continued use of the device, or can be disassembled to recover the metals in the device. A source for specifications describing bulk metals for recycling is given in Section 4.6.

3.37.2 Process Description

Recycling of electrical devices using PCB-containing dielectric oils such as transformers (see Figure 3-37) involves a combination of mechanical disassembly and chemical and thermal treatment to recover metals and, in some cases, the dielectric oil. The recycling of transformers and ballasts is a concern because older electrical equipment commonly used oils containing PCBs. Recycling a transformer involves three major steps: testing the oil for PCB content, removing the oil from the transformer, and disassembling the transformer. Once the transformer is disassembled, its components can be decontaminated, and salvageable materials can be recycled (132, 133).

3.37.3 Process Maturity

Since the initial PCB Marking and Disposal Rule in 1978, technologies have emerged to improve the efficiency of cleaning and recycling PCB oils and

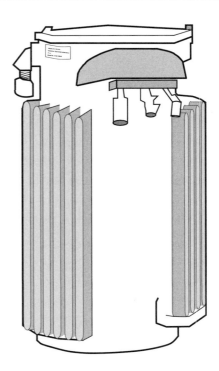

FIGURE 3-37. Example of a transfomer

transformers with less risk to the public. Methods to dechlorinate the transformer oils were in place by the early 1980s. Technologies have emerged for cleaning the oil using a mobile unit while the transformer is operating (134). For transformers that are replaced with newer, more efficient models, an estimated 90 to 95 percent of the transformer metal is recyclable (135).

3.37.4 Description of Applicable Wastes

Recovery of metals and dielectric oils is applicable to a variety of electric equipment, mainly transformers and ballast inductors. The components of the transformers include oil, tanks, cores, coils, valves, insulating materials, bushings, and other fittings, such as gauges and switches.

3.37.5 Advantages

Replacing or decontaminating transformers containing PCB fluids can reduce the costs associated with fluid testing, regulatory inspection recordkeeping, and potential spill cleanup. Metals can be salvaged from obsolete transformers. In some cases, PCB-containing dielectric oil may be dechlorinated and reused. Dechlorinated dielectric oil that is unsuitable for reuse can be treated by incineration. Incineration of transformer fluids and insulation materials from PCB transformers destroys the material and the PCBs, and the use of a metals reclamation furnace can yield even cleaner metals for recycling. Surface contamination of metal and ceramic components can sometimes be removed with the use of solvents so that the materials can be reused.

3.37.6 Disadvantages and Limitations

Thermal destruction of PCBs can create products of incomplete combustion, such as dioxins or dibenzofurans. Solvents used for surface cleaning of metal parts are distilled for reuse, but a PCB residue remains that is treated by incineration (136). Discarded materials must be carefully monitored so that Toxic Substances Control Act (TSCA) disposal requirements are not violated and no contamination of the disposal facility occurs. Materials disposed of in a landfill also continue to carry a liability, i.e., they are still the responsibility of the generator (137).

3.37.7 Operation

The actual decommissioning process starts by testing the filling oil. When handling PCB-contaminated materials, the recycling contractor implements measures to prevent, contain, and clean up spills. The oil is tested to determine the level of personal protection clothing and equipment required.

Once the oil is properly drained from the transformer, the transformer is disassembled. The transformer body, core steel, copper, aluminum, brass, and other metal components of the disassembled transformer are then accessible for recycling. Some processors clean the metal parts with solvent and transfer the metals to a smelter for recycling. Other processors use onsite incinerators, ovens, or furnaces to burn unwanted insulating materials and the adhering PCB contaminants from the transformer internals. Onsite incineration results in decontaminated metal scrap but can produce products of incomplete combustion from trace residual PCBs.

The dielectric oil typically is treated by a chemical dechlorination process. The dechlorinated oil may be either reused or destroyed by incineration.

Decontaminated materials with no commercial value, such as ceramic bushings, are sent to a landfill for disposal.

3.38 REFERENCES

When an NTIS number is cited in a reference, that document is available from:
National Technical Information Service
5285 Port Royal Road
Springfield, VA 22161
703-487-4650

1. Rogers, T.N., and G. Brant. 1989. Distillation. In: Freeman, H.M., ed. Standard handbook of hazardous waste treatment and disposal. New York, NY: McGraw-Hill. pp. 6.23-6.38.

2. California Department of Health Services. 1990. Alternative technologies for the minimization of hazardous waste. California Department of Health Services, Toxic Substances Control Program (July).

3. U.S. EPA. 1989. Project summary: Field measurements of full-scale hazardous waste treatment facilities—organic solvent wastes. EPA/600/S2-88/073. Cincinnati, OH.

4. Hertz, D.W. 1989. Reduction of solvent and arsenic wastes in the electronics industry. Hazardous Materials Management Conference and Exhibition. Oakland, CA: Association of Bay Area Governments. pp. 374-387.

5. Stanczyk, T.F. 1992. Converting wastes into reusable resources with economic value. Presented at the Government Institutes Pollution Prevention Practical Management and Compliance Strategies Training Course, Syracuse, NY (September).

6. Stoltz, S.C., and J.B. Kitte, eds. 1992. Steam: Its generation and use, 40th ed. New York, NY: Babcock & Wilcox Company.

7. Waste Tech News. 1994. Wood recycling plant under construction. Waste Tech News 6(10):4.

8. Gossman, D. 1992. The reuse of petroleum and petrochemical waste in cement kilns. Environ. Prog. 11(1):1-6.

9. U.S. EPA. 1993. Report to Congress on cement kiln dust, Vol. II: Methods and findings. EPA/530/R-94/001 (NTIS PB94-126 919). Washington, DC.

10. Bouse, E.F., Jr., and J.W. Kamas. 1988. Update on waste as kiln fuel. Rock Products 91:43-47.

11. Bouse, E.F., Jr., and J.W. Kamas. 1988. Waste as kiln fuel, part II. Rock Products 91:59-64.

12. Blumenthal, M.H. 1993. Tires. In: Lund, H.F., ed. The McGraw-Hill recycling handbook. New York, NY: McGraw-Hill.

13. PCA. 1992. An analysis of selected trace metals in cement and kiln dust. Skokie, IL: Portland Cement Association.

14. U.S. EPA. 1988. Decision criteria for recovering CERCLA wastes (draft report), Contract No. 68-01-7090, WA B-17, Task Order W61517.C6.

15. Hooper, W.B., and L.J. Jacobs, Jr. 1988. Decantation. In: Schweitzer, P.A., ed. Handbook of separation techniques for chemical engineers. New York, NY: McGraw-Hill.

16. Morgan, T.A., S.D. Richards, and W. Dimoplon. 1992. Hydrocarbon recovery from an oil refinery pitch pit. Proceedings of National Conference: Minimization and Recycling of Industrial and Hazardous Waste '92. Rockville, MD: Hazardous Materials Control Resources Institute.

17. U.S. EPA. 1991. Engineering bulletin: Thermal desorption treatment. EPA/540/2-91/008. Washington, DC.

18. U.S. EPA. 1993. Innovative treatment technologies: Semiannual status report, 5th ed. EPA/542/R-93/003. Washington, DC.

19. U.S. EPA. 1993. Low temperature thermal process for pesticides and other organic compounds. In: Tech Trends. EPA/542/N-93/007. Washington, DC.

20. U.S. EPA. 1993. Guidelines for making environmentally-sound decisions in the Superfund remedial process. Chicago, IL: Region V Waste Management Division (May).

21. Ayen, R.J., and C. Swanstrom. 1991. Development of a transportable thermal separation process. Environ. Prog. 10(3):175-181.

22. Just, S.R., and K.J. Stockwell. 1993. Comparison of the effectiveness of emerging in situ technologies and traditional ex situ treatment of solvent-contaminated soils. Emerging technologies in hazardous waste management III. American Chemical Society Symposium Series 518. Washington, DC: American Chemical Society.

23. U.S. EPA. 1992. Low temperature treatment (LT³) technology. EPA/540/AR-92/019. Washington, DC.

24. U.S. EPA. 1993. Superfund innovative technology evaluation program: Technology profiles, 6th ed. EPA/540/R-93/526. Washington, DC.

25. U.S. EPA. 1993. Applications analysis report: Resource conservation company B.E.S.T. solvent extraction technology. EPA/540/AR-92/079. Washington, DC.

26. U.S. EPA. 1990. Applications analysis report: CF systems organics extraction process, New Bedford Harbor, MA. EPA/540/A5-90/002. Washington, DC.

27. U.S. EPA. 1990. Engineering bulletin: Solvent extraction treatment. EPA/540/2-90/013. Washington, DC, and Cincinnati, OH.

28. U.S. EPA. 1991. Innovative treatment technologies: Overview and guide to information sources. EPA/540/9-91/002. Washington, DC.

29. Wisconsin Department of Natural Resources. 1993. Foundry waste beneficial reuse study. PUBL-SW-181-93. Madison, WI: Wisconsin Department of Natural Resources.

30. Ahmed, I. 1993. Use of waste materials in high-way construction. Park Ridge, NJ: Noyes Data Corporation.

31. Testa, S.M., and D.L. Patton. 1993. Soil remediation via environmentally processed asphalt. In: Hager, J.P., B.J. Hansen, J.F. Pusateri, W.P. Imrie, and V. Ramachandran, eds. Extraction and processing for the treatment and minimization of wastes. Warrendale, PA: The Minerals, Metals, and Materials Society. pp. 461-485.

32. U.S. EPA. 1990. Handbook on in situ treatment of hazardous waste-contaminated soils. EPA/540/2-90/002. Cincinnati, OH.

33. U.S. EPA. 1991. Engineering bulletin: In situ soil vapor extraction treatment. EPA/540/2-91/006. Washington, DC, and Cincinnati, OH.

34. U.S. EPA. 1991. Reference handbook: Soil vapor extraction technology. EPA/540/2-91/003. Edison, NJ.

35. U.S. EPA. 1993. Engineering forum issue: Considerations in deciding to treat contaminated unsaturated soils in situ. EPA/540/S-94/500. Washington, DC, and Cincinnati, OH.

36. Cohen, R.M., J.W. Mercer, and J. Matthews. 1993. DNAPL site evaluation. Boca Raton, FL: C.K. Smoley.

37. Superfund Week. 1993. Lead is washed from soil at TCAAP site. Superfund Week 7(44):5.

38. Villaume, J.F., P.C. Lowe, and D.F. Unites. 1983. Recovery of coal gasification wastes: An innovative approach. Proceedings of the National Water Well Association Third National Symposium on Aquifer Restoration and Ground-water Monitoring. Worthington, OH: Water Well Journal Publishing Company.

39. Heist, J.A. 1989. Freeze crystallization. In: Freeman, H.M., ed. Standard handbook of hazardous waste treatment and disposal. New York, NY: McGraw-Hill. pp. 6.133-6.143.

40. Chowdhury, J. 1988. CPI warm up to freeze concentration. Chem. Eng. 95:24-31.

41. Singh, G. 1988. Crystallization from solutions. In: Schweitzer, P.A., ed. Handbook of separation tech-

niques for chemical engineers. New York, NY: McGraw-Hill.

42. Hermann, H. 1991. Disposal of explosives from demilitarization: A case study with TNT. Presented at the 22nd International Annual Conference on ICT, Karlsruhe, Germany (July).

43. U.S. EPA. 1993. Handbook: Approaches to the remediation of federal facility sites contaminated with explosive or radioactive wastes. EPA/625/R-93/013. Washington, DC.

44. Herman, H.L. 1980. Reclamation of energetic material components from ordnance ammunition. In: Kaye, S.M. Encyclopedia of explosives and related items, PATR 2700, Vol. 9. Dover, NJ: U.S. Army Armament Research and Development Command. pp. R-146 - R-150.

45. Van Ham, N.H.A. 1991. Environmentally acceptable disposal of ammunition and explosives. Proceedings of the International Annual Conference ICT, 22nd IACIEQ, Prins Maurits Lab TNO, 2280 AA Rijswijk, The Netherlands (December).

46. U.S. Army. 1993. Conventional ammunition demilitarization master plan. Rock Island, IL: Headquarters U.S. Army Armament, Munitions, and Chemical Command.

47. Bohn, M.A., and H. Neumann. 1991. Recovery of propellant components by high-temperature high-pressure solvolysis. Presented at the 22nd International Annual Conference on ICT, Karlsruhe, Germany (July).

48. Ember, L. 1991. Cryofracture problems may prevent its use in chemical arms disposal. Chem. Eng. News 69(8):19-20.

49. Crosby, W.A., and M.E. Pinco. 1992. More power to the pop. Eng. Mining J. 193(5):28-31.

50. Torma, S.C., R.B. Rise, and A.E. Torma. 1993. Environmentally safe processing and recycling of high-energy yield materials. In: Hager, J.P., B.J. Hansen, J.F. Pusateri, W.P. Imrie, and V. Ramachandran, eds. Extraction and processing for the treatment and minimization of wastes. Warrendale, PA: The Minerals, Metals, and Materials Society. pp. 73-83.

51. HazTech News. 1994. Hydrotreating adapted to destruction of propellants, chlorinated wastes. HazTech News 9(10):74.

52. Hegberg, B.A., G.R. Brenniman, and W.H. Hallenbeck. 1991. Technologies for recycling postconsumer mixed plastics. Report No. OTT-8. University of Illinois, Center for Solid Waste Management and Research.

53. APC. 1993. Recycled plastic products source book. Washington, DC: American Plastics Council.

54. Andrews, G.D., and P.M. Subramanian. 1991. Emerging technologies in plastic recycling. ACS Symposium Series 513 (June). Washington, DC: American Chemical Society.

55. Bennett, R.A. 1991. New product applications, evaluations, and markets for products manufactured from recycled plastics and expansion of national database for plastics recycling. Technical report No. 53. Rutgers University, Center for Plastics Recycling Research.

56. Reinink, A. 1993. Chemical recycling: Back to feedstock. Plastics, Rubber, and Composites Processing and Applications 20(5):259-264.

57. Rotman, D., and E. Chynoweth. 1994. Plastics recycling: Back to fuels and feedstocks. Chem. Week 154:20-23.

58. Pearson, W. 1993. Plastics. In: Lund, H.F., ed. The McGraw-Hill recycling handbook. New York, NY: McGraw-Hill.

59. Layman, P.L. 1993. Advances in feedstock recycling offer help with plastic waste. Chem. Eng. News 71(40):11-14.

60. Randall, J.C. 1992. Chemical recycling. Mod. Plastics 69(13):54-58.

61. Shelly, S., K. Fouhy, and S. Moore. 1992. Plastics reborn. Chem. Eng. 99(7):30-35.

62. Randall, J.C. 1993. Chemical recycling. Mod. Plastics 70(12):37-38.

63. Rebeiz, K., D.W. Fowler, and D.R. Paul. 1991. Recycling plastics in polymer concrete for engineering application. Polymer Plastics Tech. Eng. 30(8):809-825.

64. Swearingen, D.L., N.C. Jackson, and K.W. Anderson. 1992. Use of recycled materials in highway construction. Report No. WA-RD 252.1. Washington State Department of Transportation.

65. SMCAA. 1992. Recycling of sheet molding compounds: The energy/environment picture. AF-184. Washington, DC: Sheet Molding Compound Automotive Alliance, Society of the Plastics Industry.

66. Sharp, L.L., and R.O. Ness, Jr. 1993. Thermal recycling of plastics. Grand Forks, ND: University of North Dakota.

67. EERC. 1994. Thermal recycling of plastics. Grand Forks, ND: University of North Dakota, Energy and Environmental Research Center.

68. Curran, P.F., and K.A. Simonsen. 1993. Gasification of mixed plastic waste. Presented at 8th Annual Recyclingplas Conference, Washington, DC (June).

69. NEESA. 1993. Precipitation of metals from ground water. Remedial Action Tech. Data Sheet Document No. 20.2-051.6. Port Hueneme, CA: Naval Energy and Environmental Support Activity.

70. U.S. EPA. 1991. Recovery of metals from sludges and wastewater. EPA/600/2-91/041. Cincinnati, OH.

71. Hickey, T., and D. Stevens. 1990. Recovery of metals from water using ion exchange. In: U.S. EPA. Proceedings of Second Forum on Innovative Hazardous Waste Treatment Technologies, Philadelphia, PA (May). EPA/540/2-90/010.

72. Brown, C.J. 1989. Ion exchange. In: Freeman, H.M., ed. Standard handbook of hazardous waste treatment and disposal. New York, NY: McGraw-Hill. pp. 6.59-6.75.

73. Ritcey, G.M., and A.W. Ashbrook. 1984. Solvent extraction principles and applications to process metallurgy. New York, NY: Elsevier.

74. ICSECT. 1988. International solvent extraction conference, Moscow, USSR (July). International Committee for Solvent Extraction Chemistry and Technology.

75. ICSECT. 1990. International solvent extraction conference, Kyoto, Japan (July). International Committee for Solvent Extraction Chemistry and Technology.

76. U.S. EPA. 1993. Project summary: Recycling nickel electroplating rinse waters by low temperature evaporation and reverse osmosis. EPA/600/SR-93/160. Cincinnati, OH.

77. Porter, M.C. 1988. Membrane filtration. In: Schweitzer, P.A., ed. Handbook of separation techniques for chemical engineers. New York, NY: McGraw-Hill.

78. MacNeil, J., and D.E. McCoy. 1989. Membrane separation technologies. In: Freeman, H.M., ed. Standard handbook of hazardous waste treatment and disposal. New York, NY: McGraw-Hill. pp. 6.91-6.106.

79. Lacey, R.E. 1988. Dialysis and electrodialysis. In: Schweitzer, P.A., ed. Handbook of separation techniques for chemical engineers. New York, NY: McGraw-Hill.

80. MSE. 1993.93. Resource recovery project technology characterization interim report. Butte, MT: U.S. Department of Energy Resource Recovery Project.

81. Delaney B.T., and R.J. Turner. 1989. Evaporation. In: Freeman, H.M., ed. Standard handbook of hazardous waste treatment and disposal. New York, NY: McGraw-Hill. pp. 7.77-7.84.

82. Grosse, D.W. 1986. A review of alternative treatment processes for metal bearing hazardous waste streams. J. Poll. Control Assoc. 36:603-614.

83. Barkay, T., R. Turner, E. Saouter, and J. Horn. 1992. Mercury biotransformations and their potential for remediation of mercury contamination. Biodegradation 3:147-159.

84. Saouter, E., R. Turner, and T. Barkay. 1994. Mercury microbial transformations and their potential for the remediation of a mercury-contaminated site. In: Means, J.L., and R.E. Hinchee, eds. Emerging technology for bioremediation of metals. Ann Arbor, MI: Lewis Publishers.

85. Horn, J.M., M. Brunke, W.D. Deckwer, and K.N. Timmis. 1992. Development of bacterial strains for the remediation of mercurial wastes. In: U.S. EPA. Arsenic and Mercury: Workshop on Removal, Recovery, Treatment, and Disposal. EPA/600/R-92/105. Washington, DC. pp. 106-109.

86. Hansen, C.L., and D.K. Stevens. 1992. Biological and physiochemical remediation of mercury-contaminated hazardous waste. In: U.S. EPA. Arsenic and Mercury: Workshop on Removal, Recovery, Treatment, and Disposal. EPA/600/R-92/105. Washington, DC. pp. 121-125.

87. Stepan, D.J., R.H. Farley, K.R. Henke, H.M. Gust, D.J. Hassett, D.S. Charlton, and C.R. Schmit. 1993. Topical report: A review of remediation technologies applicable to mercury contamination at natural gas industry sites. GRI-93/0099. Chicago, IL: Gas Research Institute; and Morgantown, WV: U.S. Department of Energy.

88. U.S. EPA. 1991. Treatment technology background document. Washington, DC.

89. Veiga, M.M., and J.A. Meech. 1993. Heuristic approach to mercury pollution in the Amazon. In: Hager, J.P., B.J. Hansen, J.F. Pusateri, W.P. Imrie, and V. Ramachandran, eds. Extraction and processing for the treatment and minimization of wastes. Warrendale, PA: The Minerals, Metals, and Materials Society. pp. 23-38.

90. U.S. EPA. 1990. Project summary: Recovery of metals using aluminum displacement. EPA/600/S2-90/032. Cincinnati, OH.

91. DeBecker, B., F. Peeters, and P. Duby. 1993. Recovery of chromium from chromate waste by electrowinning from Cr(III) solutions. In: Hager, J.P., B.J. Hansen, J.F. Pusateri, W.P. Imrie, and V. Ramachandran, eds. Extraction and processing for the treatment and minimization of wastes. Warrendale, PA: The Minerals, Metals, and Materials Society. pp. 55-72.

92. California Department of Health Services. 1989. Reducing California's metal-bearing waste streams. Technical report prepared by Jacobs Engineering Group, Inc. (August).

93. Hulbert, G., B. Fleet, M. Hess, and J. Kassirer. 1989. Technical and economic evaluation of a non-sludge, heavy metal waste reduction system at an aeroscientific corporation, Anaheim, California. Hazardous Materials Management Conference and Exhibition.

Oakland, CA: Association of Bay Area Governments. pp. 330-338.

94. Steward, F.A., and C.G. Ritzert. 1994. Waste minimization and recovery technologies. Metal Finishing 62nd Guidebook and Directory Issue 92(1A): 714-716.

95. Ramachandran, V., S.R. Gilbert, R.I. Cardenas, and M. Zwierzykowski. 1993. Recycling metal-bearing hazardous waste. In: Hager, J.P., B.J. Hansen, J.F. Pusateri, W.P. Imrie, and V. Ramachandran, eds. Extraction and processing for the treatment and minimization of wastes. Warrendale, PA: The Minerals, Metals, and Materials Society. pp. 131-141.

96. Barth, E.F., M.L. Taylor, J.A. Wentz, and S. Giti-Pour. 1994. Extraction, recovery, and immobilization of chromium from contaminated soils. Presented at the 49th Purdue University Industrial Waste Conference, Lafayette, IN (May).

97. Paff, S.W., and B. Bosilovich. 1993. Remediation of lead-contaminated Superfund sites using secondary lead smelting, soil washing, and other technologies. In: Hager, J.P., B.J. Hansen, J.F. Pusateri, W.P. Imrie, and V. Ramachandran, eds. Extraction and processing for the treatment and minimization of wastes. Warrendale, PA: The Minerals, Metals, and Materials Society. pp. 181-200.

98. Superfund Week. 1993. Mittelhauser wins design. Superfund Week 7(50):3.

99. DuGuay, T. 1993. Reclaim metals to clean up soils. Soils (March 26), pp. 28-33.

100. Lee, A.Y., A.M. Wethington, M.G. Gorman, and V.R. Miller. 1994. Treatment of lead-contaminated soil and battery wastes from Superfund sites. Superfund XIV. Rockville, MD: Hazardous Materials Control Resources Institute.

101. Prengaman, R.D., and H. McDonald. 1990. RSR's full scale plant to electrowin lead from battery scrap. In: Mackey, T.S., and R.D. Prengaman, eds. Lead-zinc '90. Warrendale, PA: The Minerals, Metals, and Materials Society.

102. Bishop, J., and M. Melody. 1993. Inorganics treatment and recovery. HazMat World 6(2):21-30.

103. Crnojevich, R., E.I. Wiewiorowski, L.R. Tinnin, and A.B. Case. 1990. Recycling chromium-aluminum wastes from aluminum finishing operations. J. Metals 42(10):42-45.

104. U.S. Air Force. 1990. Inorganic waste vitrification. Tech TIP (newsletter). Dayton, OH: Wright-Patterson Air Force Base, Joint Technology Applications Office.

105. Queneau, P.B., L.D. May, and D.E. Cregar. 1991. Application of slag technology to recycling of solid wastes. Presented at the 1991 Incineration Conference, Knoxville, TN (May).

106. Hazardous Waste Consultant. 1990. Electroplating wastes recycled by new vitrification process. Hazardous Waste Consultant 8(4):1-4 - 1-6.

107. Temini, M., A. Ait-Mokhtar, J.P. Camps, and M. Laquerbe. 1991. The use of fly ash in the clay products with cement and lime, obtained through extrusion. In: Goumans, J.J.J.M., H.A. van der Sloot, and T.G. Aalbers, eds. Waste materials in construction. Studies in environmental science 48. New York, NY: Elsevier. pp. 451-458.

108. Ali, M., T.J. Larsen, L.D. Shen, and W.F. Chang. 1992. Cement stabilized incinerator ash for use in masonry bricks. Cement Industry Solutions to Waste Management: Proceedings of the First International Conference, Calgary, Alberta, Canada (October). pp. 211-234.

109. Malolepszy, J., W. Brylicki, and J. Deja. 1991. The granulated foundry slag as a valuable raw material in the concrete and lime-sand brick production. In: Goumans, J.J.J.M., H.A. van der Sloot, and T.G. Aalbers, eds. Waste materials in construction. Studies in environmental science 48. New York, NY: Elsevier. pp. 475-478.

110. U.S. EPA. 1992. Potential reuse of petroleum-contaminated soil: A directory of permitted recycling facilities. EPA/600/R-92/096. Washington, DC.

111. Melody, M. 1993. Pollution prevention notebook: Greening soil. HazMat World 6(10):31.

112. Faase, R.W.M., J.H.J. Manhoudt, and E. Kwint. 1991. Power concrete. In: Goumans, J.J.J.M., H.A. van der Sloot, and T.G. Aalbers, eds. Waste materials in construction. Studies in environmental science 48. New York, NY: Elsevier. pp. 15-432.

113. Wainwright, P.J., and P. Robery. 1991. Production and properties of sintered incinerator residues as aggregate for concrete. In: Goumans, J.J.J.M., H.A. van der Sloot, and T.G. Aalbers, eds. Waste materials in construction. Studies in environmental science 48. New York, NY: Elsevier. pp. 425-432.

114. Bounds, C.O., and J.F. Pusateri. 1990. EAF dust processing in the gas-fired flame reactor process. In: Mackey, T.S., and R.D. Prengaman, eds. Lead-zinc '90. The Minerals, Metals, and Materials Society.

115. Hanewald, R.H., W.A. Munson, and D.L. Schweyer. 1992. Processing EAF dusts and other nickel-chromium waste materials pyrometallurgically at INMETCO. Minerals and Metallurgical Processing 9(4):169-173.

116. James, S.E., and C.O. Bounds. 1990. Recycling lead and cadmium, as well as zinc, from EAF dust. In: Mackey, T.S., and R.D. Prengaman, eds. Lead-zinc '90. Warrendale, PA: The Minerals, Metals, and Materials Society.

117. Krukowski, J. 1993. New hazardous waste solutions. Poll. Eng. 25(10):30-32.

118. Queneau, P.B., B.J. Hansen, and D.E. Spiller. 1993. Recycling lead and zinc in the United States. The Fourth International Hydrometallurgy Symposium, Salt Lake City, UT (August).

119. Versar Inc. 1988. Decision criteria for recovering CERCLA wastes. Draft report prepared for U.S. EPA Office of Emergency and Remedial Response. Springfield, VA: Versar Inc.

120. Collins, R.E., and L. Luckevich. 1992. Portland cement in resource recovery and waste treatment. Cement Industry Solutions to Waste Management: Proceedings of the First International Conference, Calgary, Alberta, Canada (October). pp. 325-331.

121. Chintis, M. 1992. Recovery of mercury D-009 and U-151 waste from soil using proven physical and gravimetric methods. In: U.S. EPA. Arsenic and Mercury: Workshop on Removal, Recovery, Treatment, and Disposal. EPA/600/R-92/105. Washington, DC.

122. Perry, R.H., and C.H. Chilton. 1984. Chemical engineers' handbook, 6th ed. New York, NY: McGraw-Hill.

123. Wills, B.A. 1985. Mineral processing technology, 3rd ed. New York, NY: Pergamon Press.

124. Lawrence, B. 1992. High vacuum mercury retort recovery still for processing EPA D-009 hazardous waste. In: U.S. EPA. Arsenic and Mercury: Workshop on Removal, Recovery, Treatment, and Disposal. EPA/600/R-92/105. Washington, DC. pp. 113-116.

125. Queneau, P.B., and L.A. Smith. 1994. U.S. mercury recyclers provide expanded process capabilities. HazMat World 7(2):31-34.

126. Weyand, T.E. 1992. Thermal treatment of mercury contaminated soils. In: Charlton, D.S., and J.A. Harju, eds. Topical report: Proceedings of the Workshop on Mercury Contamination at Natural Gas Industry Sites. GRI-92/0214. Chicago, IL: Gas Research Institute.

127. U.S. Bureau of Mines. 1993. Mineral industry surveys: Mercury 1992. Washington, DC: U.S. Department of the Interior.

128. Chemical Engineering. 1993. Tests set for mercury cleanup technologies. Chem. Eng. 100(3):50.

129. Buckley, S., N. Rouse, and D. Snoonian. 1993. The debris rule: Identification as an ARAR and cost impact of implementation at a federal facility Superfund site. Proceedings of Superfund XIV, Vol. I. Rockville, MD: Hazardous Materials Control Resources Institute. pp. 477-482.

130. U.S. EPA. 1994. Guide to cleaner technologies: Organic coating removal. EPA/625/R-93/015. Cincinnati, OH.

131. Cutler, T. 1993. Dry ice (CO_2) blaster use in battery/plating site removal. Proceedings of Superfund XIV, Vol. II. Rockville, MD: Hazardous Materials Control Resources Institute. pp. 1045-1047.

132. Dougherty, J. 1993. Transformer decommissioning: planning helps prevent liability-related problems. Electrical World 207(3):46, 56, 58.

133. HazMat World. 1994. Recycling PCB ballasts eliminates liability, risk. HazMat World 7(5):85.

134. Coghlan, A. 1993. Mobile cleaner sucks up PCBs. New Scientist 139(188-6):18.

135. Kump, R. 1993. Why should the steel industry do anything with their PCB transformers? IEEE Transactions on Industry Applications 29(5):841-845.

136. Laskin, D. 1993. What's hot and what's not: An overview of new PCB equipment and disposal technology and what plant audits are revealing. Superfund XIV Conference and Exhibition Proceedings, Vol. II. Rockville, MD: Hazardous Materials Control Resources Institute. pp. 1,411-1,415.

137. Kelly, J., and R. Stebbins. 1993. PCB regulations and procedures for risk management including PCB cleanup policy and procedures. IEEE Transactions on Industry Applications 29(4):708-715.

CHAPTER

PRODUCT QUALITY SPECIFICATIONS

Recycled materials are commodities in a competitive marketplace. Recycling is most successful where the supplier is aware of the needs of the end user. The end user is best served by a reliable supply of material conforming to an established specification. Given the variability of wastes, uniformity seldom is achieved but should be approached as closely as possible.

The volume of material available influences its re cycling potential. Larger volumes of uniform material generally are more desirable. In a few special cases, however, the waste material inventory can exceed the short-term demand. Sudden appearance of a large supply in a small market depresses the value of the potentially recyclable material.

A few waste types may be profitable to recycle. Aluminum and copper metal demolition debris can be recycled profitably if sufficient volumes of clean material are available. In most cases, significant onsite processing is needed, or a processor requires a fee to accept waste as a feedstock to an offsite recycling process.

4.1   FEED MATERIAL TO PETROLEUM REFINING

Petroleum hydrocarbons recovered at Superfund or Resource Conservation and Recovery Act (RCRA) Corrective Action sites by nonaqueous-phase liquid (NAPL) pumping (Section 3.9), thermal desorption (Section 3.5), solvent extraction (Section 3.6), or other processes often require additional pro cessing to produce a marketable petroleum product. When materials are suitable, a conventional refinery can efficiently carry out processing. This section outlines some of the key properties for determining the suitability of recovered petroleum for upgrading by conventional refinery separation processes.

The main consideration for successful refinery distillation (Section 3.1) is the difference in volatility of the components. A low boiling point mixture can be distilled at low temperature to reduce the complexity and cost of the still. When the component to be recovered is much more volatile than the contaminants, distillation can be accomplished with simple equipment.

The thermal properties (heat capacity, heat of vaporization, thermal conductivity, and heat transfer coefficient) of the material also are important. Low heat capacity and heat of vaporization indicate that low heat input is required to affect the distillation. High thermal conductivity and high heat transfer coefficient indicate that heating the waste can be accomplished with relative ease.

The physical properties must be compatible with pumping and heating the waste. A tendency to produce foam, indicated by a high surface tension, is undesirable. A waste containing high concentrations of suspended solids can form a dense, viscous sludge and clog distillation columns. High-viscosity materials are difficult to process. Organics that tend to form polymers can polymerize, clogging the column or coating the heat transfer surfaces (1).

73

To simplify distillation and maximize product quality, the waste solvent streams should be segregated as much as possible. Separating chlorinated from nonchlorinated and aliphatic from aromatic solvents is particularly beneficial (2).

4.2 ORGANIC CHEMICALS

Organic liquids and gases can be recovered from soils by processes such as thermal desorption (Section 3.5), solvent extraction (Section 3.6), or NAPL pumping (Section 3.9), or can be produced from solid materials by chemolysis (Section 3.15) or thermolysis (Section 3.17). These organic fluids are then marketed as feedstock for refineries or chemical plants. Onsite recovery of petroleum from RCRA K wastes is occurring more frequently (3, 4). For onsite recovery of plant wastes, the source of the material is known, so less documentation is needed to gain plant acceptance of the recovered feedstock.

Use of materials recovered from offsite sources raises concerns for chemical processors. Small quantities of certain impurities can poison catalysts, increase corrosion, or generate acidic off-gas. Because even small amounts of the impurities are potentially damaging, satisfactory demonstration of acceptable quality requires sophisticated sampling and analysis (5). Particularly undesirable contaminants include metals and chlorine (6). Refineries and chemical processors also require low suspended solids levels and ash content (7).

4.3 THERMOPLASTIC PARTICULATE

Thermoplastic particulate can be re-extruded into new products (Section 3.14). American Society for Testing and Materials (ASTM) Standard Guide D 5033, "The Development of Standards Relating to the Proper Use of Recycled Plastics," provides definitions of terms, describes four general types of plastic recycling, and outlines factors important in developing standards for recycling plastic. The Standard Guide notes that, unless an existing standard specifically restricts the use of recycled plastic based on performance standards, recycled plastic can be used as feedstock. Specifically mentioning recycled plastic in the specification is unnecessary. ASTM Standard Specification D 5033, "Polyethylene Plastics Molding and Extrusion Materials From Recycled Postconsumer (HDPE) Sources," defines

and specifies recycled postconsumer high-density polyethlene (HDPE) chips or pellets for molding and extrusion.

The polymer type and purity control the value of the particulate. The best candidates for reuse are single polymer types containing no impurities. Mixtures of different types of polymers, polymers containing coloring, solid additives, or impurities are much less valuable.

The properties vary substantially among the seven major polymer categories—polyethylene-terephthalate (PET), high-density polyethylene (HDPE), polyvinyl chloride (PVC), low-density polyethylene (LDPE), polypropylene (PP), polystyrene (PS), and all others. Even within a single category, polymers can have significantly different properties. For example, HDPE polymers with different molecular weights have different properties and applications. The low-molecular-weight, low-viscosity, injection-molded base cup for soft drink bottles is not interchangeable with the high-molecular-weight, high-viscosity material used in milk bottles (8). Achieving a finer separation of resin types increases the value of the recycled thermoplastic.

The price for granulated plastic typically is higher to adjust for the cost of granulation. However, users often accept only baled plastics to enable verification of impurity levels. The typical desired levels for nonplastic contaminants are either no metals or less than 3 percent nonplastic (9).

4.4 RUBBER PARTICULATE

Tires and similar rubber goods are composed mainly of polymers, carbon black, and softeners. The softeners are primarily aromatic hydrocarbons. The typical composition of a tire casing is 83 percent total carbon, 7 percent hydrogen, 2.5 percent oxygen, 1.2 percent sulfur, 0.3 percent nitrogen, and 6 percent ash (10).

Of the 278,000,000 tires discarded in 1990, about 34.5 percent were recycled, with the reuse options being retreading (13.7 percent), energy recovery (9.4 percent), fabricated products (4.3 percent), export (4.3 percent), asphalt (0.9 percent), and miscellaneous (1.9 percent) (10). Some energy recovery applications consume whole tires. Tires also can be cut into shapes or strips to make sandals, floor mats, washers, insulators, or dock bumpers (11).

For most reuse options, however, the tires must first be reduced to particulate. Particulate may be used to fabricate new rubber products or to make asphalt, or particulate can be burned to produce energy. The success of recycling and most appropriate reuse applications depends on the particle size and particle size distribution; the strength, elasticity, impurity content, and other properties; and the cost of the particulate produced.

Rubber particulate can be used to fabricate athletic field surfaces, carpet underlayment, parking bumpers, and railroad crossing beds (Section 3.16). Fine particulate rubber for these applications is produced by mechanical grinding with an abrasive or by cryogenically fracturing the material after cooling it in liquid nitrogen. Steel or fabric is separated from the rubber fragments by magnetic and/or gravity separation. The quality of the reclaimed rubber is lower than that of newly manufactured rubber because aging and treatment in the recycling process reduce elasticity.

In asphalt concrete, ground tire rubber replaces some of the aggregate in asphalt (Section 3.16). Section 1038 of the Intermodal Surface Transportation Efficiency Act of 1991 (Public Law #102-240) includes provisions to increase the number of tires used in asphalt for highway pavement.

Discarded tires are shredded to 2-in. and smaller chips for use as a fuel (Sections 3.2 and 3.3). The typical heating value is 32,500 kJ/kg (14,000 Btu/lb) for whole tire chips or 36,000 kJ/kg (15,500 Btu/lb) for steel-free tire chips. When used as fuel in cement kilns, iron from the tire belts and beads supplements the iron required for cement making. For other furnace types, the wire pieces from belts and beads are undesirable; the wires clog furnace feed equipment and generate ash. With the exception of cement kilns, most furnaces that use shredded tire fuels require dewired rubber particulate. Processing to remove the metals is commercially feasible but increases the cost of tire-derived fuel particulate.

4.5 FUELS FOR ENERGY RECOVERY

Substituting waste materials for fuel is an approach frequently applied to recover value from the waste (Sections 3.2 and 3.3). The ideal energy recovery fuel should be as similar to conventional fuels as possible. Most conventional boilers are fueled with coal, oil, or gas. ASTM D-396, "Standard Specification for Fuel Oils," divides fuel oils into grades based on suitability for specific burner types. The specification places limiting values on properties such as flashpoint, pour point, water and sediment content, carbon residue content, ash content, vaporization characteristics, viscosity, density, corrosivity, and nitrogen content. ASTM D-388, "Classification of Coal by Rank," describes classifications of coal based on factors such as carbon content, gross heat content, and agglomeration characteristics.

A high proportion of carbon and hydrogen present as organic compounds, low water content, and low ash content are the ideal conditions for a fuel material. High water content wastes heat due to the energy removed in the combustion gas by the heated water. High ash content increases the complexity of bottom ash and fly ash handling in the boiler. Specific impurities bring other possible complications. The presence of halogenated solvents in the fuel is highly undesirable because the combustion process produces acidic vapors.

Various impurities can volatilize, increasing the complexity of air pollution control requirements. High concentrations of volatile metals (arsenic, cadmium, lead, selenium, and mercury) may cause excessive concentrations of these metals to enter the combustion gases. Nonvolatile metals remain in the ash and may cause unacceptable levels of leachable metals. Halides, nitrogen, phosphorus, or sulfur can react to form corrosive gasses. Halides in the waste can combine with metals such as lead, nickel, and silver, forming volatile metal halides (12). Highly toxic chlorinated organics such as polychlorinated biphenyls (PCBs) or pesticides require high combustion temperatures and high destruction efficiencies (13).

The viscosity of liquid fuels is an important physical parameter. Liquid waste must be amenable to atomization at acceptable pressures. Liquids with a dynamic viscosity less than 10,000 standard Saybolt units (SSU) are considered pumpable. The optimum viscosity for atomization is about 750 SSU (13).

Pulp and paper making require significant energy input. To conserve resources, heat typically is supplied by onsite boilers burning wood waste (hog fuel). Hog fuel shows substantial variations in heat content and moisture. Because the hog fuel boilers are designed to feed solid wood and operate with varying quality feed, they can use solid waste materials such as broken pallets or rubber particulate more easily than conventional utility boilers can(10).

Off-specification production batches or outdated explosives contain chemical energy that can be recovered. For example, the approximate heating values of trinitrotoluene (TNT) and research department explosive (RDX) are 15,000 to 9,000 kJ/kg (6,460 and 3,900 Btu/lb), respectively (14). The use of energetic materials as fuel raises three main issues.

- The reactivity of the energetic materials must be accounted for in the design of fuel-handling and burning systems. Either the material must be dissolved in fuel oil to eliminate the possibility of explosion or the furnace must be designed to contain the largest possible explosion. TNT and RDX have low solubility in fuel oil and typically are dissolved in a solvent, such as toluene, before being mixed with fuel oil (14).

- Energetic materials contain more bound nitrogen than typical fuels, increasing the quantity of nitrogen oxide (NO_x) generated during combustion. The furnace off-gas treatment system may require special provisions to curtail or treat NO_x.

- Explosives dissolved in fuel oil increase viscosity. As discussed above, viscosity is a key parameter in the selection and design of the fuel oil atomizing nozzle. Up to limits imposed by reactivity, TNT does not significantly increase the viscosity of No. 2 fuel oil. A viscosity increase due to the addition of TNT to No. 6 fuel oil, however, is significant and may be more limiting than reactivity constraints (14).

4.6 METALS FOR REUSE

Site cleanup activities such as storage tank removal, building demolition (Section 3.36), and transformer disassembly (Section 3.37) can produce bulk metals for reuse. Iron, steel, aluminum, and copper shapes can be recycled through existing scrap recovery channels. The Institute of Scrap Recycling Industries (15) provides guideline specifications for nonferrous scrap, including copper, brass, bronze, lead, zinc, aluminum, magnesium, nickel, copper-nickel alloys, and stainless steel. These guidelines describe minimum metal content,

maximum impurity levels, density, surface contamination, and other physical conditions defining various grades of metal scrap. For example, more than 30 categories of aluminum scrap are described.

For recycling of bulk metal from demolition, surface contaminants such as welds, rust, scale, paint, or coatings generally are acceptable. Hazardous contaminants must be removed, however (16). Mixtures of alloy types of the same metal also reduce value. Paradoxically, very expensive specialty stainless steels or aluminum alloys may be less valuable in the recycling market than low-alloy materials; the added alloying metals are viewed as impurities in the recycling process.

4.7 METAL-CONTAINING SLUDGE OR SLAG FOR FEED TO SECONDARY SMELTERS

Waste materials at Superfund or RCRA Corrective Action sites may contain a sufficiently high concentration of metals to be suitable for processing in a smelter (Section 3.31). Lower-concentration wastes may be amenable to processing by another method (e.g., physical separation [Section 3.33] or chemical leaching [Section 3.29]/precipitation [Section 3.18]) that would produce a smaller volume of residual with a metals content high enough to warrant smelting. The metal types and concentrations, matrix properties, and impurities control the suitability of wastes for processing in secondary smelters. The typically desired minimum concentrations of six metals for secondary smelters are indicated in Table 4-1.

The waste matrix also may contribute constituents needed to form slag. The main slag-form-

TABLE 4-1
Approximate Feed Concentration Requirements for Secondary Smelters (13, 17, 18)

Metal	Approximate Minimum Concentration for Pyrometallurgical Recovery
Cadmium	2%
Chromium	5%
Copper	30%
Lead	55%
Nickel	1.3%
Zinc	8%

ing components are silica, iron, and calcium. A waste matrix with high thermal conductivity is desirable. High thermal conductivity indicates that the matrix can be heated more quickly and uniformly (19).

Impurities that volatilize or react to form volatile products increase the complexity and expense of off-gas treatment. Examples include mercury, arsenic, nitrates, sulfates, sulfides, phosphates, and halides. Mercury and arsenic are amenable to pyrometallurgical processing but require special off-gas treatment provisions (1). Except for arsenic in lead battery alloys, arsenic- and mercury-containing materials are incompatible with most existing secondary smelters in the United States.

Pyrometallurgical processing relies on partitioning of different metals to vapor, slag, and molten metal phases to form purified products. Impurities that partition to the same phase as the target metal are undesirable. Incompatibilities are process specific. Volatile metals such as arsenic and antimony tend to contaminate the zinc oxide product fumed from a waelz kiln. Silver and bismuth are difficult to separate from lead in secondary refining because they tend to remain in the metal rather than partition to the slag.

Alkaline metals such as sodium and potassium decrease the viscosity and increase the corrosiveness of slag formed in a pyrometallurgical reactor. Excessive levels of alkaline metals increase the difficulty in controlling slag properties and may cause the slag to damage the reactor lining (20).

4.8 WASTE FEED TO HYDROMETALLURGICAL PROCESSING

Hydrometallurgical processes, such as acid leaching (Section 3.29), precipitation (Section 3.18), reverse osmosis (Section 3.21), or ion exchange (Section 3.19) are not highly selective. The difficulty of recovering products from mixed metal wastes increases as the number of metal contaminants present increases. Segregating spent processing solutions greatly enhances their compatibility with recycling processes. The waste stream should not be mixed or diluted. The highest concentration of metal possible should be maintained (21).

Typical parameters of interest for hydrometallurgical processing include calcium, cadmium, chromium, copper, iron, mercury, magnesium, nickel, phosphorus, lead, tin, zinc, organic content, color, smell, acid insolubles, moisture, cyanide, and filtration rate. Organics interfere with chemical leaching and many of the solution processing techniques used to recover metals from the leaching solution. Organic contaminants in wastes to be treated by hydrometallurgical methods should, therefore, be minimized (19). The speciation of the metal contaminants in the waste is an important factor. The valence state and counter ion affect the ability to dissolve the metal. Chelating agents or metal complexing anions can greatly increase the difficulty of recovering metals from solutions (22). Wastes with high pH and high alkalinity are difficult to treat by acid leaching due to the high consumption of acid by the reserve alkalinity.

4.9 HIGH-VALUE CERAMIC PRODUCTS

Some waste streams can be treated at high temperature to produce valuable ceramic products (Section 3.30). For example, vitrified waste can be fritted to make abrasives, cast from a melt, or formed and sintered into products such as bricks or architectural dimension stone. The waste also may be converted to a frit for use as feed material in the manufacture of ceramics. Ceramic materials are products manufactured by high-temperature treatment of raw materials of mainly earthy origin. The main components of ceramics are silicon, silica, and/or silicate. The variety of possible products includes:

- *Structural clay products,* which include burned clay products such as brick, roof tile, and ceramic tile and pipe.

- *Refractories,* which include special materials for high-temperature applications such as kiln-lining bricks, high-temperature insulation materials, and castable refractories.

- *Abrasives,* which include particulate material (along with any supporting materials and binders) used for cutting, grinding, or polishing; common abrasives are fused alumina, silicon carbide, silica, alumina, and emery.

- *Architectural products,* which include decorative and structural ceramics such as brick, blocks, patio stones, wall and floor tile, art pottery, and chemical and electrical porcelain.

- *Glass products,* which include vitreous silicate products such as window glass, container glass, and glass fibers.

- *Porcelain enamel products,* which include products with a ceramic coating on a metal substrate such as sink and bathroom fixtures, architectural panels, and specialty heat and chemical-resistant equipment.

A variety of organizations publish specifications for ceramic feed materials or products. Example sources of specifications for ceramic products include ASTM, the American National Standards Institute, the U.S. Navy (MIL-A-22262(SH) for sandblasting media), and the Steel Structures Painting Council (SSPC-XAB1X for mineral and slag abrasives).

4.10   INORGANIC FEED TO CEMENT KILNS

Manufacture of hydraulic cement, a conventional building material, offers possibilities for recycling of contaminated waste materials (Section 3.32). Making hydraulic cement requires a significant input of energy and raw materials. Opportunities exist for input of nonhazardous metals-contaminated solids to cement kilns. Of particular interest to the recycling of metals-contaminated waste is the need for silica, iron, and alumina.

In raw material grinding, the input materials are ground so that 75 to 90 percent of the material passes through a 0.074-mm (2.9 in.) (200-mesh) sieve. The grinding may be done either wet or dry. In wet milling, water is added with the mill feed to produce a slurry containing about 65 percent solids.

Specifications for limestone feed for cement making require that the calcium carbonate ($CaCO_3$) content be greater than 75 percent and the magnesium carbonate ($MgCO_3$) content be less than 3 percent. Because the raw materials must be finely ground, chert nodules or coarse quartz grains are undesirable (23).

ASTM specifies five basic types of Portland cement. Type I is intended for use when the special properties of the other types are not required. Type IA is for the same uses as Type I where air entrainment is desired. Air entrainment is a technique to improve freeze/thaw resistance of the concrete and reduce the mix viscosity without increasing water addition. Type II also is a general-use cement but of-fers increased sulfate resistance and lower heat generation. Type IIA is similar to Type II but is intended for use where air entrainment is desired. Type III is formulated to maximize early strength production. Type IIIA is the air entrainment version of Type III. Type IV is intended for use where the heat generation must be minimized. Type V is for use when sulfate resistance is desired. The main constituents of Portland cement typically are tricalcium silicate (C_3S), dicalcium silicate (C_2S), tricalcium aluminate (C_3A), and tetracalcium aluminoferrite (C_4AF).

The U.S. production of Portland and masonry cement in 1991 was about 70,000,000 metric tons (77,000,000 tons). Portland cement makes up 96 percent of the total U.S. cement output. Types I and II account for about 92 percent of Portland cement production.

4.11   CEMENT SUBSTITUTE

Use as a cement supplement or substitute is a viable option for some fly ash and slag wastes (Section 3.7). Fly ash is used in large quantities to stabilize sulfate sludge, and it can replace cement in construction applications. Construction applications require selection of fly ash that is low in sulfate impurity and consists of small, generally spherical particles. Spherical particles act to reduce the mix viscosity and thus allow preparation of concrete with less water addition (25). In mass concrete pours, excessive temperature increases may occur due to the heat of hydration released as the concrete sets. Replacement of some cement by a pozzolan can reduce the generation of this heat. Fly ash addition also can be valuable in reducing heat generation (26).

Slag cooled with sufficient speed to retain a largely vitreous structure exhibits cementitious properties when hydroxide activators are present. Blast furnace slag from iron production is the most commonly used slag pozzolan, but other types are used if they contain limited quantities of free calcium oxide (CaO) or magnesium oxide (MgO). Free alkaline earth oxides may reduce strength due to delayed hydration. Steel-making slags are reportedly poor candidates due to their high calcium content (26). However, magnesium slags are reported to be good candidates (27). In waste solidification/stabilization treatment, replacing some cement with blast furnace slag provides reducing power to help hold metals in a less mobile chemical state (28).

Portland cement is the most commonly used type of hydraulic cement. Recent environmental regulations and increased energy costs have increased the cost of cement making, which has increased the attractiveness of substituting slag pozzolans for conventional cement (29). Blended cements are available and may be used to reduce costs or for special purposes.

The U.S. Environmental Protection Agency (EPA) has developed guidelines to assist agencies in the procurement of cements and concretes that contain fly ash (40 CFR Part 249). Subpart B of this guide describes the development of the guide and contract specifications to allow use of cement containing fly ash, as well as provides specific recommendations for material specifications. For cement, these recommendations are:

- ASTM C 595—Standard Specification for Blended Hydraulic Cements.
- Federal Specification SS-S-1960/4B—Cement, Hydraulic, Blended.
- ASTM C 150—Standard Specification for Portland Cement and Federal Specification SS-C-1960/ General (appropriate when fly ash is used as a raw material in the production of cement).

For concrete, these recommendations are:

- ASTM C 618—Standard Specification for Fly Ash and Raw Calcined Natural Pozzolan for Use as a Mineral Admixture in Portland Cement Concrete.
- Federal Specification SS-C-1960/5A—Pozzolan for Use in Portland Cement Concrete.
- ASTM C 311—Standard Methods of Sampling and Testing Fly Ash and Natural Pozzolans for Use as a Mineral Admixture in Portland Cement Concrete.

Subpart C describes recommended approaches to bidding and price analysis. Subpart D describes recommended certification procedures including measurement, documentation, and quality control.

4.12 AGGREGATE AND BULK CONSTRUCTION MATERIALS

Sand and slag wastes can be used directly for var-

ious construction purposes (Section 3.7). Other inorganic wastes can be vitrified to produce rock-like materials (Section 3.30). The main requirements in using waste materials as aggregates or bulk materials are regulatory acceptance, customer acceptance, and performance. The waste material must meet the required leach resistance criteria and provide some useful function in the product; the end use should not be simply disposal in another form. Even if regulatory requirements are met, construction companies and local citizens are reluctant to accept the use of waste materials. Therefore, leach resistance and durability testing may be required beyond those specified in the regulations; the reused waste should meet the performance requirements of new materials. Some sources of information on performance of aggregate and construction materials are outlined below.

Aggregate is a mineral product from natural or manufactured sources used in concrete making. The specifications for fine and coarse aggregate are described in ASTM 33. The important features of aggregate are size grading; freedom from deleterious materials such as clay lumps, friable particles, and organic materials; and soundness.

The alkali reactivity of the cement and aggregate is an important factor in selecting an aggregate. The concern is reaction of an alkali with the aggregate, causing a volume increase and/or loss of concrete strength. The alkali causing the reaction usually is the calcium hydroxide released as the cement cures. In some cases, however, the alkali may come from external sources, such as ground water. There are two basic types of alkali- aggregate reactions:

- Reaction of alkali with siliceous rocks or glasses.
- Reaction of alkali with dolomite in some carbonate rocks.

Some waste slags can exhibit excessive reactivity. For example, four zinc smelter slag samples tested by Oklahoma State University were found to be unsuitable as aggregate for Portland cement because of excessive expansion during curing caused by alkali aggregate reactions (30).

One of several tests can determine the alkali activity of a potential aggregate, depending on the type of aggregate to be tested. The applicable tests or guides are ASTM C 227, "Test Method for Potential Alkali Reactivity of Cement-Aggregate Combinations (Mortar-Bar Method)"; ASTM C 289, "Potential Reactivity of Aggregates (Chemical Method)"; ASTM C 295, "Petrographic Examination of Aggregates for Concrete"; ASTM C 342, "Standard Test Method for Potential Volume Change of Cement-Aggregate Combinations"; and ASTM C 586, "Potential Alkali Reactivity of Carbonate Rocks for Concrete Aggregates (Rock Cylinder Method)." Guidance for selecting the appropriate test method is given in ASTM C 33, "Standard Specification for Concrete Aggregates."

Coarse aggregate for bituminous paving mixtures is specified in ASTM D 692. This specification covers crushed stone, crushed hydraulic-cement concrete, crushed blast-furnace slag, and crushed gravel for use in bituminous paving mixtures specified in ASTM D 3515 or D 4215. Air-cooled blast-furnace slag is required to have a compacted density of not less than 1120 kg/m^3 (70 lb/ft^3) when sizes number 57 [25 to 4.75 mm (1 to 0.19 in.)] or 8 [9.5 to 2.36 mm (0.38 to 0.09 in.)] (ASTM C 448) are tested. Additional guidance on polishing characteristics, soundness, and degradation is given.

The ASTM and the American Association of State Highway and Transportation Officials (AASHTO) are the main national organizations setting specifications regarding crushed stone for use in construction. However, states or localities develop many specifications for construction aggregates based on their specific needs. Most common specifications control size grades, soundness, shape, abrasion resistance, porosity, chemical compatibility, and content of soft particles. Due to the skid resistance imparted to road surfaces when blast furnace or steel furnace slag is used as the aggregate, many state agencies specify that slag aggregate for asphalt be applied to roads with high traffic volume (29).

The American Railroad Engineering Association sets standards for railroad ballast. The general characteristics of a good ballast material are strength, toughness, durability, stability, drainability, cleanability, workability, and resistance to deformation.

Limestone for lime manufacture should contain more than 90 percent CaCO$_3$, less than 5 percent MgCO$_3$, and less than 3 percent other impurities. Feed to vertical limestone kilns should be 12.7 to 20.3 cm (5 to 8 in.) in size. A size range of 9.5 to 64 mm (8^3/$_8$ to 2^1/$_2$ in.) is acceptable for limestone feed to rotary kilns.

Specifications for limestone or dolomite for fluxing metallurgical processes depend on the type of ore to be processed and the intended end use of the slag. A silica content of less than 2 to 5 percent, a magnesia content in the range of 4 to 15 percent, and a sulfur content of less than 0.1 percent are typical for fluxstone specifications.

Limestone or dolomite for glass manufacture should contain more than 98 percent CaCO$_3$ or MgCO$_3$, respectively. The iron limit for glass-making input typically is 0.05 to 0.02 percent.

The American Water Works Association established specification B100-94, "Standards for Filtering Materials," for particulate used in filtration operations. The specification describes criteria affecting the acceptability of filtration media such as particle shape, specific gravity, effective grain size and uniformity, acid soluble impurity content, and radioactive and heavy metal content.

4.13 REFERENCES

1. U.S. EPA. 1991. Treatment technology background document. Washington, DC.

2. California Department of Health Services. 1990. Alternative technologies for the minimization of hazardous waste. California Department of Health Services, Toxic Substances Control Program (July).

3. Horne, B., and Z.A. Jan. 1994. Hazardous waste recycling of MGP site by HT-6 high temperature thermal distillation. Proceedings of Superfund XIV Conference and Exhibition, Washington, DC (November 30 to December 2, 1993). Rockville, MD: Hazardous Materials Control Resources Institute. pp. 438-444.

4. Miller, B.H. 1993. Thermal desorption experience in treating refinery wastes to BDAT standards. Incineration Conference Proceedings, Knoxville, TN (May).

5. Randall, J.C. 1992. Chemical recycling. Mod. Plastics 69(13):54-58.

6. Reinink, A. 1993. Chemical recycling: Back to feedstock. Plastics, Rubber, and Composites Processing and Applications 20(5):259-264.

7. Morgan, T.A., S.D. Richards, and W. Dimoplon. 1992. Hydrocarbon recovery from an oil refinery pitch pit. Proceedings of National Conference: Minimization and Recycling of Industrial and Hazardous Waste '92. Rockville, MD: Hazardous Materials Control Resources Institute.

8. Pearson, W. 1993. Plastics. In: Lund, H.F., ed. The McGraw-Hill recycling handbook. New York, NY: McGraw-Hill.

9. Hegberg, B.A., G.R. Brenniman, and W.H. Hallenbeck. 1991. Technologies for recycling post-consumer mixed plastics. Report No. OTT-8. University of Illinois, Center for Solid Waste Management and Research.

10. Blumenthal, M.H. 1993. Tires. In: Lund, H.F., ed. The McGraw-Hill recycling handbook. New York, NY: McGraw-Hill.

11. Carless, J. 1992. Taking out the trash. Washington, DC: Island Press.

12. U.S. EPA. 1992. Superfund engineering issue: Considerations for evaluating the impact of metals partitioning during the incineration of contaminated soils from Superfund sites. EPA/540/S-92/014. Washington, DC.

13. Versar Inc. 1988. Decision criteria for recovering CERCLA wastes. Draft report prepared for U.S. EPA Office of Emergency and Remedial Response. Springfield, VA: Versar Inc.

14. Myler, C.A., W.M. Bradshaw, and M.G. Cosmos. 1991. Use of waste energetic materials as a fuel supplement in utility boilers. J. Haz. Mat. 26(3):333-341.

15. Institute of Scrap Recycling Industries. 1991. Scrap specifications circular. 1991 guidelines for nonferrous scrap: NF-91. Washington, DC.

16. von Stein, E.L. 1993. Construction and demolition debris. In: Lund, H.F., ed. The McGraw-Hill recycling handbook. New York, NY: McGraw-Hill.

17. Hanewald, R.H., W.A. Munson, and D.L. Schweyer. 1992. Processing EAF dusts and other nickel-chromium waste materials pyrometallurgically at INMETCO. Minerals and Metallurgical Processing 9(4):169-173.

18. James, S.E., and C.O. Bounds. 1990. Recycling lead and cadmium, as well as zinc, from EAF dust. In: Mackey, T.S., and R.D. Prengaman, eds. Lead-zinc '90. Warrendale, PA: The Minerals, Metals, and Materials Society.

19. U.S. EPA. 1991. Recovery of metals from sludges and wastewater. EPA/600/2-91/041. Cincinnati, OH.

20. Queneau, P.B., L.D. May, and D.E. Cregar. 1991. Application of slag technology to recycling of solid wastes. Presented at the 1991 Incineration Conference, Knoxville, TN (May).

21. Edelstein, P. 1993. Printed-circuit-board manufacturer maximizes recycling opportunities. HazMat World 6(2):24.

22. St. Clair, J.D., W.B. Bolden, E.B. Keough, R.A. Pease, N.F. Massouda, N.C. Scrivner, and J.M. Williams. 1993. Removal of nickel from a complex chemical process waste. In: Hager, J.P., B.J. Hansen, J.F. Pusateri, W.P. Imrie, and V. Ramachandran, eds. Extraction and processing for the treatment and minimization of wastes. Warrendale, PA: The Minerals, Metals, and Materials Society. pp. 299-321.

23. Tepordei, V.V. 1992. Crushed stone annual report 1990. Washington, DC: U.S. Department of the Interior, Bureau of Mines (April).

24. Johnson, W. 1992. Cement annual report 1990. Washington, DC: U.S. Department of the Interior, Bureau of Mines.

25. Horiuchi, S., T. Odawara, and H. Takiwaki. 1991. Coal fly ash slurries for backfilling. In: Goumans, J.J.J.M., H.A. van der Sloot, and T.G. Aalbers, eds. Waste materials in construction. Studies in environmental science 48. New York, NY: Elsevier. pp. 545-552.

26. Popovic, K., N. Kamenic, B. Tkalcic-Ciboci, and V. Soukup. 1991. Technical experience in the use of industrial waste for building materials production and environmental impact. In: Goumans, J.J.J.M., H.A. van der Sloot, and T.G. Aalbers, eds. Waste materials in construction. Studies in environmental science 48. New York, NY: Elsevier. pp. 479-490.

27. Courtial, M., R. Cabrillac, and R. Duval. 1991. Feasibility of the manufacturing of building materials from magnesium slag. In: Goumans, J.J.J.M., H.A. van der Sloot, and T.G. Aalbers, eds. Waste materials in construction. Studies in environmental science 48. New York, NY: Elsevier. pp. 491-498.

28. Bostick, W.D., J.L. Shoemaker, R.L. Fellows, R.D. Spence, T.M. Gilliam, E.W. McDaniel, and B.S. Evans-Brown. 1988. Blast furnace slag: Cement blends for the immobilization of technetium-containing wastes. K/QT-203. Oak Ridge, TN: Oak Ridge Gaseous Diffusion Plant.

29. Solomon, C.C. 1992. Slag-iron and steel annual report 1990. Washington, DC: U.S. Department of the Interior, Bureau of Mines (April).

30. U.S. EPA. 1990. Report to Congress on special wastes from mineral processing. EPA/530/SW-90/070C. Washington, DC.

CHAPTER

5

CASE STUDIES

This chapter highlights specific case studies of successful examples of commercial recycling of complex waste materials. Eight case studies were selected to illustrate applications of a variety of recycling technologies covering a wide range of contaminant and matrix types. The case studies describe:

- Use of spent abrasive blasting media as aggregate in asphalt.

- Use of spent abrasive blasting media as a raw material for Portland cement making.

- Physical separation to recover lead particulate from soils at small-arms practice ranges.

- Processing lead-containing wastes from Superfund sites in a secondary smelter.

- A treatment train for recovery of petroleum from an oily sludge.

- Solvent recovery using small onsite distillation units.

- Thermal desorption to clean soil for reuse.

- Pumping to recover coal tar liquids.

Each case study includes sections on site and waste description, technology description, recycling benefits, economic characteristics, and limitations.

These case studies show real-world examples of ways to overcome the challenges of implementing recycling technologies, as well as demonstrate the application of treatment trains to produce useful products from complex waste mixtures.

5.1 RECYCLING SPENT ABRASIVE BLASTING MEDIA INTO ASPHALT CONCRETE

Reuse as asphalt aggregate may be feasible for a wide variety of petroleum or metals-contaminated soils, slags, or sands (Section 3.7). The Naval Facilities Engineering Services Center in Port Hueneme, California, has been studying the recycling of spent abrasive blasting media (ABM), or sandblasting grit, into asphalt concrete for commercial paving purposes. The sandblasting grit is used as a "blender sand" for a portion of the fine-grained aggregate that is used to produce the asphalt concrete. This section briefly describes a case history for an ongoing "ABM-to-asphalt" recycling project in Hunters Point, California.

5.1.1 Site and Waste Description

The spent ABM at Hunters Point is composed of a 2,300-m^3 (3,000-yd^3) pile of Monterey Beach sand contaminated with small amounts of paint chips. The spent ABM was generated in shipcleaning operations conducted at Naval Station, Treasure Island, Hunters Point Annex, by Triple AAA Shipcleaning during the 1970s and 1980s. The spent ABM grades as a coarse sand and contains relatively low concentrations of metals. Average copper, zinc, lead, and chromium concentrations are 1,800, 1,100, 200, and 100 mg/kg (105, 64, 12, and 6 grains/gal), respectively. Leachable metals concentrations using the California Waste Extraction Test (WET) methodology average 140, 150, 20, and 2 mg/L (8.2, 8.8, 1.2, 0.12 grains/gal), respectively, for copper, zinc, lead, and chromium. The WET test is California's version of the U.S. Environmental Protection Agency's (EPA's) Toxicity Characteristic Leaching Procedure (TCLP). The spent ABM at Hunters Point is considered hazardous because of soluble threshold limit concentration (STLC) exceedances on the WET test for copper and lead but is not an EPA hazardous waste because it passes the TCLP.

Different types of spent ABM other than Monterey Beach sand can be recycled into asphalt concrete. A coal slag-derived ABM from shipcleaning operations has been recycled successfully in Maine. A variety of ABM products derived from both primary and secondary smelter slags also are recyclable, including copper and nickel slags.

Waste types other than spent ABM also can be recycled into asphalt concrete as a substitute for a portion of the aggregate. For example, spent foundry sand and sandy or gravelly textured soils have been successfully recycled (1, 2). Mixed colored glass is being recycled into asphalt in New Jersey, and the product has been termed "glassphalt" (3). Numerous permitted facilities recycle petroleum-contaminated soils into asphalt concrete; EPA (4) provides a directory of these facilities. Rubber from tires can be pyrolized and substituted for a portion of the bitumen in the asphalt concrete. Tire particulate can be used as aggregate in asphalt concrete (5). Also, worn-out asphalt pavement can be crushed, graded, and recycled as aggregate in asphalt concrete (6).

5.1.2 Technology Description

The ABM-to-asphalt recycling technology involves substituting the ABM for a portion of the fine-size aggregate in asphalt concrete. As long as the metal concentrations in the spent ABM are not excessively high, the metal concentrations in the asphalt concrete product should be very low, and any metals present must be physically and chemically immobilized in the asphalt binder. Typically, asphalt concrete consists of approximately 5 percent bitumen and 95 percent graded aggregate. The graded aggregate includes particles varying from fine sand to 12- to 25-mm (½- to 1-in.) gravel. Depending on the mix design and the ultimate strength requirements of the product, the fine-size particle fraction may comprise 25 to 35 percent of the asphalt con-

crete. In the ABM-to-asphalt technology demonstration at Hunters Point, an ABM concentration of 5 percent by weight of the final asphalt concrete is being used. In other words, spent ABM equals 5 percent of the asphalt concrete and approximately one-seventh to one-fifth of the normal fine fraction component of the asphalt concrete. Higher ABM contents are possible; theoretically, the entire fine fraction of the mix design could be composed of ABM. At higher ABM concentrations, however, a greater potential exists for adverse impact on product quality and/or elevated metals concentrations in the product.

ABM recycling is applicable to both cold- and hot-mix asphalt processes. At Hunters Point, the ABM is being recycled into hot-mix asphalt for normal commercial paving applications, yielding high-strength asphalt concrete for heavily used highways. ABM can be recycled into both a base coarse layer or any subsequent lifts applied to the base coarse. ABM also can be recycled into cold-mix processes, which yield a lower-grade product for road repair or lower-traffic-area applications.

5.1.3 Recycling Benefits

The spent ABM at Hunters Point is hazardous in the state of California and, if no recycling and reuse option were available, would have to be treated by stabilization/solidification and disposed of in a hazardous waste landfill. This technology makes beneficial reuse of the ABM by incorporating it into asphalt concrete, where resulting metal concentrations are low and the metals have been immobilized in the asphalt concrete matrix. Millions of tons of asphalt concrete are produced in the United States annually; therefore, there is a considerable demand for aggregate for asphalt pavement.

5.1.4 Economic Characteristics

The cost of an ABM-to-asphalt recycling project depends on a number of factors, particularly:

- Tippage rate charged by the asphalt plant.
- Distance from the generator to the asphalt plant, which affects transportation costs.
- Required amount of planning, regulatory interactions, reporting, and program management.

In addition, the following factors affect cost to a lesser degree:

- Analytical fees for chemical and physical analyses of asphalt test cores to show compliance with any regulatory or institutional requirements.

- ABM pretreatment, such as screening and debris disposal.

In the Hunters Point project, the tippage rate charged by the asphalt plant is $44 per metric ton ($40 per ton) of ABM recycled. The overall cost is approximately $155 per metric ton ($140 per ton), including significant costs for transportation to the asphalt plant, regulatory compliance, and analytical testing of core specimens produced in the laboratory prior to full-scale recycling. In general, the recycling costs decrease on a per-ton basis with increasing amounts of spent ABM recycled. The following ranges are typical for most projects:

Amount of ABM (Tons)	Estimated Costs of Recycling (per Ton)
500–1,500	$125–$175
1,500–3,000	$100–$150
3,000–6,000	$50–$100

Therefore, the ABM-to-asphalt recycling approach is economically beneficial for both the asphalt plant and the ABM generator. The generator pays significantly less per ton than it would for disposal in a hazardous landfill and probably less than it would cost for onsite treatment and disposal, and the asphalt plant receives payment for a raw material for which it ordinarily has to pay.

5.1.5 Limitations

The asphalt recycling approach is viable for only certain types of aggregates. The aggregate must comply with both performance and environmental standards such as durability, stability, chemical resistance, biological resistance, permeability, and leachability (7). The principal limitations pertain to risk, regulatory considerations, or technical considerations pertaining to the integrity of the asphalt concrete product. For example:

- ABM-containing solvents or other particularly hazardous or toxic constituents should not be recycled in this manner.

- ABM with high metal contents (percent level or greater) may pose hazards either to workers at the asphalt plant due to dust exposure or to the public in the asphalt product because of metals leaching.

- The presence of sulfate or metallic iron is undesirable because these materials swell upon hydration of sulfates or oxidation of iron. Reduced forms of trace metals may cause similar problems, which, however, may be avoidable by recycling the ABM into a base coarse layer, where there is minimal contact with air.

- High concentrations of silt and smaller size particles are undesirable because they have poor wetting characteristics in the bitumen matrix and may generate dusts.

- Highly rounded aggregates are not compatible with good vehicular traction in the asphalt concrete product.

The chief chemist or engineer at the asphalt plant must ensure that the ABM is compatible with the production of a high-integrity asphalt concrete product.

Cognizant regulators should be contacted prior to proceeding with the recycling project. RCRA regulations discourage the land application of recycled hazardous materials (8). In most cases, special wastes or state-regulated wastes may be recyclable, subject to state or local restrictions or policies.

5.2 RECYCLING SPENT ABRASIVE BLASTING MEDIA INTO PORTLAND CEMENT

Silicate matrices containing iron or aluminum are good candidates for reuse as cement raw materials (Section 3.32). The Naval Facilities Engineering Services Center in Port Hueneme, California, along with Southwestern Portland Cement Co., Mare Island Naval Shipyard, Radian Corporation, and Battelle, has been studying the recycling of spent ABM, or "sandblasting grit," as a raw material for the manufacture of Portland Type I cement for construction purposes. The ABM is a silicate slag containing moderate levels of iron and is being used as a substitute for the iron ore that normally is used in cement manufacture. The silica and alumina in the ABM are also useful ingredients in the cement product. This section briefly describes a case history for an ongoing "ABM-to-Portland-cement" recycling project being conducted at Southwestern Portland Cement in Victorville, California.

5.2.1 Site and Waste Description

The source of the ABM is Mare Island Naval Shipyard in Vallejo, California, which generates approximately 1,800 metric tons (2,000 tons) of spent ABM per year from sandblasting submarines. The ABM recycled in this demonstration project is derived from a slag from copper smelting. The average bulk elemental composition of this slag-derived abrasive is as follows:

- Iron oxide as Fe_2O_3—23 percent
- Silica as SiO_2—45 percent
- Alumina as Al_2O_3—7 percent
- Calcium as CaO—19 percent
- Sodium as Na_2O—less than 0.2 percent
- Potassium as K_2O—less than 0.1 percent
- Magnesium as MgO—6 percent

The abrasive has a total copper concentration of approximately 0.2 percent. In addition, the ABM becomes contaminated with additional copper and other metals during sandblasting. The types and concentrations of metals depend on the types of paints and coatings being removed. Typical metal concentrations in the spent ABM recycled in this demonstration are shown below (mg/kg):

- Copper—3,120
- Barium—1,080
- Zinc—197
- Vanadium—118
- Chromium—90
- Cobalt—70
- Nickel—62
- Lead—33
- Arsenic—25

The spent ABM is considered hazardous in the state of California because of its copper content but is not a hazardous waste under Resource Conser-

vation and Recovery Act (RCRA) definitions. Consequently, this recycling demonstration is being conducted under a research and development variance issued by the California Environmental Protection Agency.

Several waste types other than spent ABM also are good candidates for recycling in this manner, particularly wastes high in alumina (such as bottom or fly ash, ceramics, and aluminum potliner) and/or iron (e.g., iron mill scale and foundry waste). Silica and calcium also are beneficial ingredients but usually are provided in sufficient quantities by the quarry rock; therefore, they are not as much in demand.

5.2.2 Technology Description

The manufacture of Portland cement includes preparation, grinding, and exact proportional mixing of mineral feedstocks, followed by heating and chemical processing in the kiln. The raw materials necessary for cement production include limestone (or another source of calcium carbonate), silica, alumina, and iron oxides that can be provided by clay, diatomaceous earth, inorganic wastes, or other sources. The feedstocks are tested for chemical and physical constituents and are mixed in exact proportions to obtain the required properties of the produced cement (see also Section 3.32).

In the more energy-saving kiln operations, the raw materials are fed through a "calciner." This process uses residual heat from the kiln and adds additional heat to begin the important calcining reaction or the dissociation of carbon dioxide from the calcium carbonate to form calcium oxide or "quicklime." Figure 5-1 presents a simplified diagram of the cement manufacturing process for a kiln equipped with a precalciner. In older cement manufacturing operations, the process is somewhat simplified and limited to a single rotating kiln in which all calcining and chemical reactions occur.

Regardless of the process, the material is passed through the rotary kiln, which heats the mixture up to 1,480°C (2,700°F). At this temperature, the calcium oxide reacts with silica and alumina to form calcium silicates and aluminates, the primary components of cement. The resulting products at the end of the kiln are the hardened nodules known as clinker. These nodules are allowed to cool, then

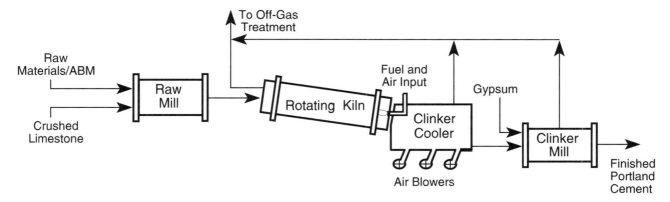

FIGURE 5-1. Abrasive blasting material and the cement-making process

finely ground and combined with gypsum to create the final product, Portland cement (9).

During the demonstration tests, ABM was introduced as approximately 1 percent of the total feedstock of the kiln, and emissions monitoring was conducted to identify any fluctuations in the air emissions concentrations from the process. The final product was then subjected to physical and chemical analysis to determine the structural integrity of the product and whether the metals are bound in the crystalline structure of the cement. The results of these tests showed that the ABM in these proportions did not significantly increase the metals content of the clinker or lead to undesirable air emissions (9).

5.2.3 Recycling Benefits

The spent ABM at Mare Island Naval Shipyard is hazardous in the state of California and, if no recycling and reuse options were available, would have to be treated by stabilization/solidification and disposed of in a hazardous waste landfill. This technology makes beneficial reuse of the ABM by incorporating it into Portland cement, where resulting metal concentrations are low and the metals are physically and chemically immobilized in the cement chemical matrix. Tens of millions of tons of Portland cement are produced in the United States annually; therefore, there is a considerable demand for iron- and aluminum-rich feedstock for cement production.

For example, 11 cement manufacturers currently operate 20 Portland cement kilns in the state of California. In 1989 alone, these operations reported the cumulative production of more than 9.4 million

metric tons (10.4 million tons) of cement clinker. Due to gaseous losses during the calcining reaction, approximately 12.2 million metric tons (13.5 million tons) of feedstock were required to generate the cement. Therefore, if only one-tenth of 1 percent of the required feedstock for each of these kilns were dedicated to recycling of metal-containing wastes, up to 12,200 metric tons (13,500 tons) of hazardous waste could be diverted from landfill disposal in just the state of California each year (9).

5.2.4 Economic Characteristics

ABM use in cement manufacturing presents a positive economic opportunity to both the waste generator and the operator of the cement kiln. In this demonstration, the total fee charged by the kiln operator has been about $215 per metric ton ($195 per ton), and approximately 3,630 metric tons (4,000 tons) of spent ABM have been recycled thus far. This fee covers a number of different costs on the part of the kiln operator, including:

- The cost of transporting the spent ABM from the generator's site in northern California to the cement plant in southern California.

- Costs incurred by the kiln operator for determining feedstock proportions and for process modifications to accommodate the waste materials.

- The cost of sampling and analyzing both the clinker and the air emissions from the stock to ensure that elevated metals concentrations are not being generated in either medium.

- Costs associated with regulatory compliance and any necessary permits or variances.

The only significant cost element not included in the $215 per metric ton ($195 per ton) figure is the cost of ABM screening and debris disposal, which was borne by the shipyard and probably amounted to less than $11 per metric ton ($10 per ton).

Up to the point of the initiation of this recycling project, Mare Island Naval Shipyard was spending approximately $730 per metric ton ($660 per ton) to manage this waste stream, including characterization, transportation, and disposal in a hazardous waste landfill (including any treatment required by the landfill operator). Therefore, the cost savings to the generator are obvious and significant, and the kiln operator is being paid to take a raw material for which the cement plant usually has to pay.

5.2.5 Limitations

Recycling into Portland cement is applicable to only certain types of wastes, based on chemical composition, contaminant levels, and other criteria (10, 11):

- Aluminum, iron, and sometimes silica are the primary constituents that the kiln operator needs to purchase to supplement the naturally occurring concentrations in the quarry rock. Ores typically comprise 40 to 50 percent by weight of these constituents. Therefore, waste materials should contain at least 20 percent or more of these constituents to be attractive substitutes for the ore materials.

- Combustion to heat the raw materials and decomposition reactions during formation of cement clinker generate large volumes of off-gas, which must be controlled and cleaned.

- Elevated concentrations of volatiles such as sodium, potassium, sulfur, chlorine, magnesium, and barium are adverse to cement production and can lead to problems such as premature oxidation or the production of excess quantities of kiln dust or acidic volatiles such as hydrochloric acid. Product quality specifications for inorganic feed to cement kilns are discussed in Section 4.10. The plant chemist is the final authority on whether a given waste material is compatible with the mix design.

- Recycling operations must not create a significant risk due to elevated metals concentrations in the clinker or off-gas. Total metals concentrations in the recycled wastes should in general be less than 1 percent, and the clinker must be tested to ensure that metals present are not highly leachable. Waste with highly toxic and volatile metals such as arsenic or mercury should not be recycled in this manner.

Cognizant regulators should be contacted prior to proceeding with the recycling project. RCRA regulations discourage the land application of recycled hazardous materials (8). In most cases, special wastes or state-regulated wastes may be recyclable, subject to state or local restrictions or policies.

5.3 RECOVERING LEAD PARTICULATE FROM SMALL-ARMS PRACTICE RANGES

Between the armed services, municipalities, and private clubs, there are tens of thousands of outdoor small-arms ranges, either active or abandoned, in the United States. Small arms are pistols, rifles, and machine guns with calibers of 15 mm (0.6 in.) or less. Because of the inevitable buildup of bullets in the target and impact berms, these ranges are potential source areas for metals contamination. These sites also are excellent candidates for metals recovery and recycling because a major portion of the bullets can be easily removed by separation technologies based on differences in size and/or density (Section 3.33). The U.S. Naval Facilities Engineering Services Center in Port Hueneme, together with Battelle and the U.S. Bureau of Mines, has been studying technologies for recycling lead and other metals from small-arms impact berms. One recycling demonstration has been completed at Naval Air Station Mayport, Florida, and others have been undertaken at Camp Pendleton, California; Quantico, Virginia; and other bases.

5.3.1 Site and Waste Description

Impact berms of small arms ranges come in various sizes. The height of the berm may vary from 1.5 m (5 ft) to as high as 15 m (50 ft), and lengths may vary from 4.6 m to 1.609 km (15 ft to 1 mile). The "average" volume of a berm, based on a survey (12), is about 3,100 yd^3. The estimated "average" volume of contaminated berm soil is 627 m³ (820 yd³), as-suming a 0.9-m (3-ft) depth of contamination on the impact side of the berm.

The average mass of lead accumulated in an impact berm is about 3,190 kg (7,030 lbs) per year and can reach as high as 9,000 kg (19,850 lbs) per year. The average fraction of lead by volume in the contaminated soil is about 1 to 2 percent, but localized pockets can contain up to 30 percent or more by volume. Most of the lead is in the form of large pieces of bullets and can be separated from the berm soil using physical separation techniques such as screening. Bullet fragments typically will be retained by a 3.5-mesh (5.66-mm [0.22-in.]) screen. For example, in a test run for lead-containing soil at the Camp Pendleton site, a 3.5-mesh screen retained 8.5 percent of the 1.36 metric tons (1.5 tons) screened but collected 59.4 percent of the 143 kg (316 lbs) of the total lead contaminant. Other metals such as copper, zinc, tin, and antimony are also frequently present due to their occurrence in lead alloys and copper or brass jackets. Removing the pieces of bullets usually will not render the soil clean, however, because a proportion of the metal contamination also exists in the form of small fragments and weathering products or lead ions adsorbed onto the soil matrix. For example, in one study the total elemental lead concentration in a berm soil was 23,000 ppm, even after sieving the soil with an 80-mesh (0.117-mm [0.005-in.]) screen (12).

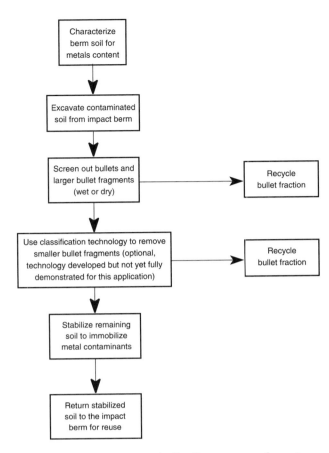

FIGURE 5-2. Recovering bullet fragments and reusing berm soil at small-arms practice ranges

5.3.2 Technology Description

A flowchart for the technology is shown in Figure 5-2. Simple screening, either dry or wet, usually can remove 80 to 90 percent or more of the bullet fragments present. Dry screening is a simple operation and is effective on most dry and coarse soils. Wet screening can separate smaller bullet fragments and, therefore, remove a greater fraction of the lead; also, wet screening is more effective on wet or clayey soils and does not generate dust. Wet screening is more complex, however, and involves peripheral equipment such as mixing tanks, pumps, thickeners, and possibly chemical additives. Also, an additional waste stream of water that may be contaminated with lead is generated and must be dealt with.

Screening usually is effective at removing the majority of the recyclable metal present; however, depending on the site, the use of classification or grav-ity concentration technology may be warranted if a significant fraction of the metal has a fine particle size that is not recoverable by screening, for example at rifle (as opposed to pistol) ranges, where bullets tend to fragment more upon impact. Because both the soil materials and the bullet fragments can have a range of particle sizes, the separation equipment should be selected based on site-specific conditions. Table 5-1 lists some typical separation processes and applicable particle-size ranges. Note, however, that advanced separation technology is not fully demonstrated at small range sites and can be very time-consuming and expensive to implement.

The lead-rich fractions from processing the impact berm soil can be recycled to a primary or secondary lead smelter. Primary smelters (e.g., Doe Run in Boss, Missouri, or ASARCO in Helena, Montana) accept small range soils with relatively low lead levels (several percent by weight), whereas

TABLE 5-1
Particle Size Range for Application
of Separation Techniques (13,14)

Separation Process	Particle Size Range
Screening	
Dry screen	>3,000 μm
Wet screen	150 μm
Hydrodynamic classifiers	
Elutriator	>50 μm
Hydrocyclone	5–150 μm
Mechanical classifier	5–100 μm
Gravity concentrators	
Jig	>150 μm
Spiral concentrator	75–3,000 μm
Shaking table	75–3,000 μm
Bartles-Mozley table	5–100 μm
Froth flotation	5–500 μm

secondary lead smelters prefer 40 percent lead or more.

Even after recovering the majority of the elemental lead and other metals from the berm soil, there is a good possibility that the berm soil will fail the TCLP test and therefore require further treatment. If this is the case, the soil can be treated using stabilization/solidification (S/S) technology to immobilize the remaining metals. The treated soil can then be returned to the berm for reuse or be disposed of in accordance with applicable regulations.

5.3.3 Recycling Benefits

This recycling process treats the impact berm as a mineral deposit to be mined for lead and other metals. As indicated above, a majority of the lead can be removed easily by screening, resulting in a concentrate that is acceptable for recycling to a primary or secondary lead smelter. Removal of lead simultaneously greatly reduces the lead content of the berm soil and the risk to human or ecological receptors. After treatment by S/S technology, the berm soil can be returned to the impact berm for future use. The presence of S/S chemicals in the treated berm soil acts as a "chemical buffer" to inhibit any future leaching of lead or other metals that are introduced into the berm during future use.

5.3.4 Economic Characteristics

Dry screening is relatively inexpensive and usually can be performed for $11 to $22 per metric ton ($10 to $20 per ton). Wet screening and classification or gravity concentration technology have higher capital and operating costs because of the auxiliary equipment and may produce a secondary wastewater stream that requires additional cost to treat. S/S treatment averages approximately $165 per metric ton ($150 per ton), depending on the cost of the treatment chemicals and the tonnage of soil to be stabilized (15). The lead-rich concentrate can be sent to a recycler, who will charge a tippage fee of between $130 to $440 per metric ton ($120 to $400 per ton) provided the concentrate contains at least 10 percent lead by weight. If the lead content is high enough (typically over 70 percent) and concentrations of other metallic impurities such as copper and zinc are low enough (less than 5 to 10 percent), the recycler pays up to $155 per metric ton ($140 per ton) to receive the material. A list of secondary lead smelters in the United States is provided in Table 5-2. Several primary lead smelters also are operating in the United States and Canada. Another significant cost element for recycling the lead concentrate is shipping, which averages approximately $0.17 per metric ton per loaded kilometer ($0.25 per ton per loaded mile).

Therefore, the overall cost of this impact berm recycling/remediation technology for the typical size impact berm, assuming the stabilized berm soil can be reused in the impact berm, averages $110 to $275 per metric ton ($100 to $250 per ton), depending on the volume of soil to be stabilized and whether the recycler pays or charges for taking the lead bullets. This compares favorably with disposal in a hazardous waste landfill.

5.3.5 Limitations

Bullet concentrate containing less than 10 percent lead is not attractive to a secondary lead smelter but can probably be recycled to a primary smelter, assuming transportation costs are not excessive. Copper and zinc are not desirable but should not cause the soil to be unacceptable as long as the concentrations of these metals are below the lead concentration. Soils with excessive sodium, potassium, sulfur, and/or chloride can pose complications due

TABLE 5-2
U.S. Secondary Lead Smelters
as of November 1993
(adapted from Queneau and Troutman [16])

Smelter Location	Year Built	Approximate Capacity mtpya	Furnace Type
Ponchatoula, LA	1987	8,000	BF-SRK
Boss, MO	1991	65,000	REV (paste) SRK (metal)
Lyon Station, PA	1964	54,000	REV-BF
Muncie, IN	1989	70,000	REV-BF
Reading, PA	1972	65,000	REV-BF
College Grove, TN	1953	10,000	BF
Eagan, MN	1948	55,000	REV-BF
Tampa, FL	1952	18,000	BF
Columbus, GA	1964	22,000	BF
Frisco, TX	1978	55,000	REV-BF
Los Angeles, CA	1981	90,000	REV-BF
Rossville, TN	1979	9,000	BF
City of Industry, CA	1950	110,000	REV
Indianapolis, IN	1972	110,000	REV-BF
Wallkill, NY	1972	70,000	REV
Troy, AL	1969	110,000	REV
Baton Rouge, LA	1960	70,000	REV-BF
Forest City, MO	1978	27,000	BF
Total secondary lead smelting capacity		1,023,000	

BF = blast furnace; REV = reverberatory furnace; SRK = short rotary kiln
aAs lead metal

5.4 PROCESSING OF SUPERFUND WASTES IN A SECONDARY LEAD SMELTER

The Center for Hazardous Materials Research and an industrial partner are testing methods to recover lead from Superfund site wastes by processing in conventional secondary lead smelters (Section 3.31).

5.4.1 Site and Waste Description

The demonstration has researched the potential for lead recovery from a variety of waste matrices containing less than 50 percent lead. The typically desired concentration for feed to a secondary smelter is 50 percent. The waste types tested in the program include battery cases with 3 to 10 percent lead; lead drosses, residues, and debris containing 30 to 40 percent lead; and lead paint removal debris containing 1 percent lead (17).

5.4.2 Technology Description

As shown in Figure 5-3, a typical secondary lead smelter upgrades lead-bearing feed in a two-step process. Most of the feed to secondary lead smelters is from recycled lead products, mainly spent lead-acid batteries. In 1992, about 75 percent of U.S. lead production came from recycle of old scrap (16).

In the conventional lead smelting process, batteries are broken up and processed by gravity separation. Poly- propylene case material often is recycled to make new battery cases. Older hard rubber cases are used as fuel in the reverberatory furnace. The lead plates and lead-containing paste are processed in the smelter.

The first stage of pyrometallurgical treatment in the reverberatory furnace selectively reduces the feed material to produce a relatively pure soft lead metal product and a lead oxide slag containing about 60 to 70 percent lead. The slag also contains battery alloy elements such as antimony, arsenic, and tin, as well as impurities. Any sulfur in the feed exits the reverberatory furnace in the off-gas as sulfur dioxide (SO_2). Solids in the off-gas are removed by filtration and are returned to the furnace.

The lead-oxide-rich slag is reduced in the blast furnace to produce hard-lead bullion and waste slag. Iron and limestone are added to enhance the lead purification process. Lead metal is cast for reuse. The

to volatility in the blast furnace and off-gassing. The secondary smelter should be contacted prior to initiating the project to ensure that the lead concentrate meets the smelter's acceptance criteria.

In certain areas, the regulators may not permit the stabilized soil to be returned to the impact berm or to be disposed of on site. If this is the case and the impact berm soil must be disposed of in a hazardous landfill, then there may be little incentive to recover and recycle the lead concentrate prior to disposing of the waste in the hazardous landfill.

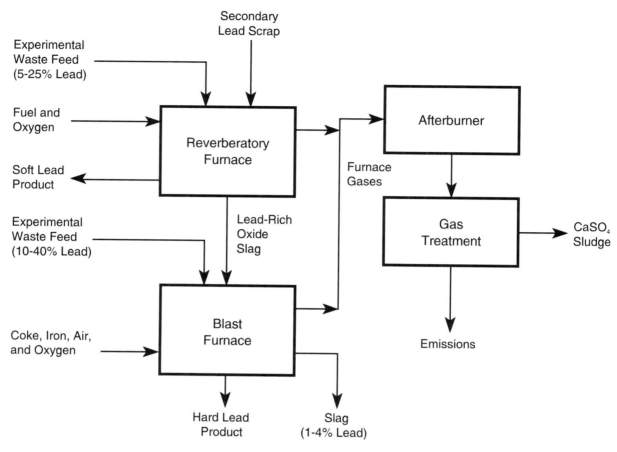

FIGURE 5-3. Processing lead wastes in a secondary smelter (18).

slag, containing about 1 to 4 percent lead, is disposed of in a RCRA-permitted landfill.

As shown in Figure 5-3, wastes containing 5 to 25 percent lead are added to the reverberatory furnace, and wastes containing 10 to 40 percent lead are added to the blast furnace. The lower grade wastes were proportioned in with normal higher-grade feed materials. Modifications were required to reduce the particle size of some wastes and to incorporate the wastes in with the normal feed materials.

5.4.3 Recycling Benefits

This recycling process removes lead from wastes at the Superfund site and returns the metal to commercial use. Although some lead remains in the slag, the total quantity of lead entering landfills is reduced. Increasing the reuse of old scrap decreases the demand for lead ore, thus reducing the environmental degradation due to mining and smelting of primary lead ores.

5.4.4 Economic Characteristics

The test program demonstrated that the costs for treating materials containing 10 to 40 percent lead range from $165 to $275 per metric ton ($150 to $250 per ton). These costs are competitive with S/S treatment and landfill disposal (19). Treatment of material with lower lead concentration typically is not economical for a secondary smelter (17).

Transportation costs can be significant when large volumes of material must be moved over long distances. Table 5-2 indicates the locations, capacities, and processing systems of secondary lead smelters in the United States.

5.4.5 Limitations

Wastes containing a low lead concentration and a high proportion of silica reduce the pollution prevention benefit due to increased slag production. Slag from the blast furnace contains 1 to 4 percent lead. Wastes with little or no combustible content and lead concentrations less than about 10 percent will provide little or no net lead recovery.

Wastes containing significant concentrations of chlorides may increase the concentration of lead in the flue dust due to formation of volatile $PbCl_4$.

5.5 TREATMENT TRAIN FOR RECOVERY OF PETROLEUM FROM OILY SLUDGE

A combination of decanting (Section 3.4) and thermal desorption (Section 3.5) is being applied on a commercial scale to recover petroleum products from oily sludge wastes at refineries (20, 21). These techniques can be applied to recovery of petroleum or solvents from wastes at Superfund or RCRA Corrective Action sites.

5.5.1 Site and Waste Description

The oil recycling system is a permanently installed capital addition at a petroleum refinery. The material being processed is listed RCRA waste with codes in the range of K048 to K052. These are source-specific wastes from petroleum refining, including dissolved air flotation sludge, oil emulsion, heat exchanger cleaning sludge, API separator sludge, and tank bottoms. The land disposal restriction contaminants of concern in these waste streams include benzene, toluene, ethylbenzene, xylene, anthracene, benz(a)anthracene, benzo(a)pyrene, *bis*(2-ethylhexyl)phthalate, naphthalene, phenanthrene, pyrene, cresol, and phenol. The boiling points of the organic constituents of the waste range from about 80°C to 390°C (175°F to 730°F).

5.5.2 Technology Description

The oil recovery process uses physical decanting (Section 3.4) and thermal desorption (Section 3.5) in sequence to recover the petroleum. The listed K wastes are collected in feed tanks, where diatomaceous earth may be added if a bulking agent is needed. The mixed oil, water, and solids stream is then passed through a centrifuge.

Treatment in the centrifuge produces a mixed oil/water stream and a sludge stream. The oil/water stream is separated in an oil/water separator. The oil returns to the refinery, and the water is treated for discharge. The sludge stream is treated by drying and thermal desorption to recover additional petroleum and reduce the levels of organic contaminants in the treated residue to below concentrations specified by the treatment standards of the Land Disposal Restrictions.

Thermal treatment of the solids from the centrifuge occurs in two stages. The solids are first processed in a steam-heated screw dryer operating about 80°C to 110°C (175°F to 230°F) to remove water. The dried solids are then processed in a hot oil-heated screw thermal desorption unit to vaporize the organic contaminants. The operating temperature of the thermal desorber is 425°C to 450°C (800°F to 840°F), and the residence time is about 1 hr.

Off-gas from the dryer, desorber, and the enclosed conveyor systems connecting the process units is collected and treated with a water scrubber followed by final cleanup with granular activated carbon. The scrubber water is routed to the oil/water separator to recover the desorbed petroleum products.

5.5.3 Recycling Benefits

The recycling process treats about 2,360 to 3,150 wet metric tons per month (2,600 to 3,470 wet tons per month) of oily sludge waste to recover petroleum for return to the refinery. The output from the thermal desorber is about 90.8 metric tons (100 tons) of dry solid. The organic content is reduced to meet the requirements of the Land Disposal Restrictions. The dry solids are further treated in a permitted on-site land farm to remove the trace levels of organics remaining.

5.5.4 Economic Characteristics

The process is estimated to save $40,000 per month compared with reuse of the sludge for energy recovery, or $100,000 per month compared with incineration.

5.5.5 Limitations

The desorbed organic is mainly high-boiling-point material with a high viscosity. For the application described in the case study, the high viscosity is not a problem. Because the process operates at a refinery, the recovered material is added to the refinery input stream. In other applications, a local use for the heavy oil, such as at an asphalt plant, would be needed or the recovered material would have to be shipped to a refinery for additional processing.

5.6 SOLVENT RECOVERY BY ONSITE DISTILLATION

Solvent distillation is a widely applied technology for recovering a reusable product from wastes containing significant amounts of solvent (Section 3.1). Although this case study describes a process application, the approach also could be applied to cleanup of organic liquids in abandoned tanks or drums, or to liquids obtained by nonaqueous phase liquid (NAPL) pumping (Section 3.9), thermal desorption (Section 3.5), or solvent extraction (Section 3.6).

5.6.1 Site and Waste Description

Two types of small, onsite distillation units were tested. An atmospheric batch still was tested on spent methyl ethyl ketone (MEK) at a site where MEK is being used to clean the spray painting lines between colors. The recycled solvent was reused for the same purpose, and the residue was shipped off as hazardous waste. The vacuum still was tested on spent methylene chloride (MC) at a site where wires and cables are manufactured. The MC is being used for cold (immersion) cleaning of wires and cables to remove ink markings.

In appearance and color, the spent samples were vastly different from the recycled and virgin samples. The recycled samples were relatively similar in appearance and color to the virgin samples, giving the first indication that contamination had been reduced during recycling. The specific gravity value of the recycled samples fell between those of the spent and virgin samples. Absorbance measurements indicated sharp differences between the spent solvent and the recycled solvent. There was little difference between absorbance measurements on the recycled and virgin samples. Therefore, appearance, color, specific gravity, and absorbance could serve as quick indicators of solvent quality for onsite operators.

Nonvolatile matter (contamination) accounted for nearly 7 percent of the spent MEK sample. This was reduced to approximately 0.002 percent in the recycled sample. The conductivity and acidity values of the recycled samples fell between those of the spent and virgin samples, indicating some improvement in these parameters. The water content increased from

approximately 1.9 percent in the spent sample to approximately 5.5 percent in the recycled samples. This increase indicates that water contamination present in the spent solvent transfers to the distillate. However, the fact that the volume of the distillate is roughly 30 percent lower than the total volume of the initial spent batch explains only a part of this increase in water concentration. The remaining water must have entered the recycled solvent during the recycling process itself, possibly due to a slight leakage from the water-cooled condenser that was worn out from several months of use.

The purity of the recycled MEK sample showed a substantial improvement from the spent sample, increasing from 78 percent to about 85 percent. The large decrease in nonvolatile matter during recycling (discussed above) accounts for most of this increase in purity. Of the 15 percent impurity in the recycled sample, 5 percent is water, as discussed above. The remaining 10 percent impurity is probably due to the co-distilling out of paint thinner solvents (proprietary blends) present in the spent solvent.

5.6.2 Technology Description

This case study evaluated two different technologies for recovering and reusing waste solvent on site. The two technologies tested were atmospheric batch distillation and vacuum heat-pump distillation. In each technology category, a specific unit offered by a specific manufacturer was tested. However, other variations of these units (with varying capabilities) are available from several vendors.

5.6.2.1 Atmospheric Batch Distillation

This is the simplest technology available for recovering liquid spent solvents. Units that can distill as little as 19 L (5 gal) or as much as 208 L (55 gal) per batch are available. Some of these units can be modified to operate under vacuum for higher-boiling solvents (150°C [300°F] or higher). Contaminant components that have lower boiling points than the solvent or that form an azeotrope with the solvent cannot be separated (without fractionation features) and may end up in the distillate. The distillation residue, which often is a relatively small fraction of the spent solvent, is then disposed of as hazardous waste.

The unit has several safety features, including explosion-proof design. A water flow switch/interlock ensures that the unit shuts off if water supply to the condenser is interrupted. The still has to be installed in an area with explosion-proof electrical components. Generally, solvent users already have a flammables storage area where the still can be installed. Insurers occasionally may require additional safety features, such as explosion-proof roofing.

5.6.2.2 *Continuous Vacuum Distillation*

As shown in Figure 5-4, the configuration is similar to that of a conventional vacuum distillation system except that the pump, in addition to drawing a vacuum, functions as a heat pump. No external heating or cooling is applied. The heat pump generates a vacuum for distillation and compresses vapors for condensation. The model used in the testing is suitable for solvents with boiling points up to 80°C (175°F). The spent solvent is continuously sucked into the evaporator by a special filling valve. The vacuum drawn generates vapors, which are sucked into the heat pump, compressed, and sent to the condenser. The still operating temperature stabilizes automatically according to the specific solvent and the ambient temperature. The condenser surrounds the

evaporator, allowing heat exchange between the cool spent solvent and the warm condensing vapors.

The heat pump is a single-stage rotary vacuum pump, modified to operate in a solvent atmosphere. The pump oil is a type that is insoluble in solvent. Solvent vapor entering the pump is kept free of solid and liquid impurities by a vapor filter and condensate trap. An overflow protection device guards against foaming in the evaporator by releasing the vacuum. A continuous distillate is produced and can be collected in a clean tank or drum. The residue at the bottom of the still can be intermittently drained for spent solvents containing less than 5 percent solids. For spent solvents with a higher solids content, continuous draining by means of a discharge pump may be necessary.

5.6.3 Recycling Benefits

Described below is the waste reduction achieved at the test site by the two distillation technologies at the respective sites. Through recycling, large volumes of spent solvent waste were reduced to small volumes of distillation residue, which is disposed of as RCRA hazardous waste. In the case of the vacuum unit, a very small sidestream of used oil is generated through a routine oil change on the vacuum pump. This oil is combined with other waste oil generated on the site and disposed of according to state regulations for used oil disposal. According to the manufacturer, air emissions due to the recycling process itself are largely avoidable, provided that the operating procedures recommended by the manufacturer are followed. This site has modified the unit to process faster, however, and this results in some air emissions (825 L/yr [218 gal/yr]) due to incomplete condensation of the vapors.

Both MEK and MC are hazardous chemicals listed on the Toxics Release Inventory (TRI). These solvents are also on EPA's list of 17 chemicals targeted for 50 percent reduction by 1995.

5.6.4 Economic Characteristics

The economic evaluation compares the costs of onsite solvent recycling versus purchase of new solvent and disposal of spent solvent.

The atmospheric distillation case study indicated that annual solvent use decreased from 3,330 to 927 L (880 to 245 gal), and that the volume of sol-

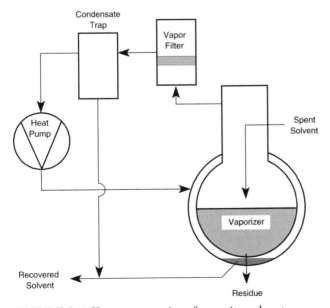

FIGURE 5-4. Vacuum vaporizer for onsite solvent distillation

vent residue requiring disposal decreased from 3,400 to 980 L (900 to 260 gal). Solvent recycling resulted in savings of $10,011 per year. The purchase price of the atmospheric batch unit is $12,995. A detailed calculation based on worksheets provided in the Facility Pollution Prevention Guide (22) indicated a payback period of less than 2 years.

The vacuum distillation unit reduced solvent use from 11,350 to 950 L (3,000 to 250 gal) per year. The solvent volume requiring disposal dropped from 11,350 to 515 L (3,000 to 136 gal). The savings due to recycling amounted to $18,283 per year. The purchase price of the vacuum unit is $23,500 for the explosion-proof model. The payback period for this unit also was less than 2 years.

5.6.5 Limitations

Solvent distillation is a versatile technology that can be used in a number of different applications. The main limitation of this technology comes into effect if the solvent to be recovered and the contaminants have similar boiling characteristics (vapor pressures). When the liquids have similar boiling characteristics, simple distillation equipment will not allow a good separation. The waste solvent would then have to be taken to an offsite location, where facilities for fractional distillation are available.

5.7 THERMAL DESORPTION TO TREAT AND REUSE OILY SAND

A thermal desorption (Section 3.5) unit is processing abrasive sand contaminated with diesel fuel to remove the oily contaminant and allow reuse of the treated sand as fill for new construction at the site (23).

5.7.1 Site and Waste Description

The waste matrix is an abrasive silica sand located at the Los Angeles Port Authority San Pedro Harbor. The sand is contaminated with marine diesel fuel that has petroleum hydrocarbon levels as high as 30,000 mg/kg.

5.7.2 Technology Description

Sand is first screened to remove large debris, then fed through a counterflow kiln where it is heated to 427°C (800°F). Organics are vaporized and treated in a thermal oxidizer. The hydrocarbons also act as supplemental fuel in the oxidizer. A cyclone and baghouse filter system collects the fine particulates. Collected particulate is returned to the kiln for treatment. A lining of refractory ceramic tiles has been provided in the unit's ducting and cyclone to protect against the abrasive sand.

5.7.3 Recycling Benefits

The thermal desorption process will produce 272,340 metric tons (300,000 tons) of clean sand. The sand will be used as fill during the construction of a new container storage facility, thus avoiding the environmental impact and cost of disposal. Residual petroleum hydrocarbon levels following treatment average 71 mg/kg, as measured by EPA SW-846 method 8015M. The site requirement called for reducing hydrocarbons to 1,000 mg/kg or below. To allow reuse at the site, the total polycyclic aromatic hydrocarbon (PAH) concentration must be below 1 mg/kg. The thermal desorption process is removing PAH compounds to nondetectable concentrations.

5.7.4 Economic Characteristics

Cleanup and reuse of the soil avoids the costs for purchase and transport to the site of clean fill and the cost for disposal of a large volume of sand. The ceramic lining requires inspection and minor refurbishment for each 90,780 metric tons (100,000 tons) of sand processed.

5.7.5 Limitations

Thermal treatment of the desorbed organic generates some NO_x. Small amounts of particulate also are released, even with the off-gas treatment filtration system.

5.8 PUMPING TO RECOVER NONAQUEOUS- PHASE LIQUIDS

Free product can be recovered from in situ formations by pumping (Section 3.9) followed by decanting (Section 3.4). Nine months of operation with an in situ pumping system recovered more than 26,500 L (7,000 gal) of virtually water-free, high-Btu-content product (24).

5.8.1 Site and Waste Description

The site was a former coal gasification plant located in the borough of Stroudsburg, Pennsylvania. Coal tar, produced as a byproduct of coal gasification, had been discharged through trenches, wells, and the ground surface at the site. The site investigation indicated that up to 6.8 million L (1.8 million gal) of free coal tar had been distributed over about 3.2 hectares (8 acres). The coal tar extended from the surface down to a silty sand deposit but was not able to penetrate that layer. An accumulation of up to 132,000 L (35,000 gal) of nearly pure coal tar was held in a stratigraphic depression.

5.8.2 Technology Description

Coal tar was recovered using a well cluster 76 cm (30 in.) in diameter installed at the deepest point of the depression. The cluster consisted of four wells 15 cm (6 in.) in diameter screened only in the coal tar layer and one central well 10 cm (4 in.) in diameter screened over its entire length.

Initially, coal tar was recovered by pumping from the four wells at a slow rate. The center well was used for monitoring. About 380 L (100 gal) per day of nearly pure product was recovered at the start of pumping, but the withdrawal rate declined rapidly over time.

The system was modified, as shown in Figure 5-5, to increase the product withdrawal rate. The center well was modified by installing a plug at a depth between the static ground-water and static coal tar levels. The central well was then used to remove ground water. Ground water was reinjected in a gravel-filled leaching field located about 19.8 m (65 ft) upgradient. The resulting reduction in ground-water pressure over the pumping well caused an upwelling of the coal tar. Cyclic pumping of the coal tar was used, with pump operation controlled by two conductivity sensors. One conductivity sensor was located at the maximum upwelling level and the other at the static (per pumping) coal tar level. Coal tar pumping cycled on when the coal tar level reached the upper conductivity sensor, and cycled off when pumping had lowered the coal tar level back to the static level.

5.8.3 Recycling Benefits

Nine months of pumping collected about 26,500 L (7,000 gal) of coal tar containing less than 1 percent water, with a heating value of 40,700 kJ/kg (17,500 Btu/lb). The recovered coal tar was used as supple-

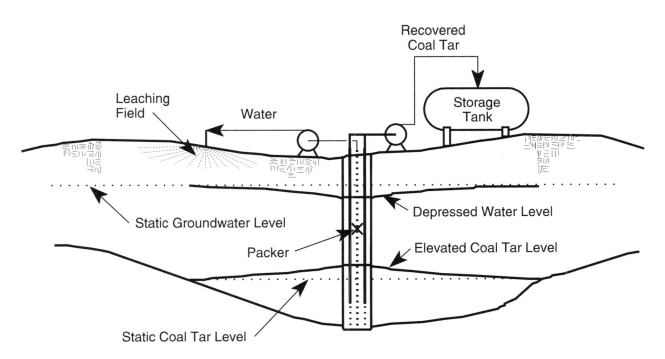

FIGURE 5-5. Coal tar recovery system (adapted from Villaume et al. [24])

mental fuel and potentially could have been used as a chemical feedstock.

5.8.4 Economic Characteristics

The coal tar recovery system was designed for unattended operations and functioned well in that mode. The major operating costs were electrical service for the pumps and rental fees for pumps and storage tanks. Maintenance requirements were minimal and involved periodic replacement of product recovery pump impellers and cracked polyvinyl chloride (PVC) piping. Waste disposal was minimal due to the essentially closed-loop operation. The operating and maintenance costs typically were $1,000 per month.

5.8.5 Limitations

The most significant problem encountered during pumping operations was attack of well equipment by the coal tar. The coal tar embrittled plastics such as PVC piping, electrical insulation, and other equipment, causing cracking.

5.9 REFERENCES

1. New England Waste Resources. 1991. Remediation saves Worcester site. New Eng. Waste Res. (March), pp. 12-13.

2. Testa, S.M., and D.L. Patton. 1992. Add zinc and lead to pavement recipe. Soils (May), pp. 22-35.

3. Monroe, C.K. 1990. Laboratory investigation 88-3/59-R-323, bituminous concrete glassphalt route 37-86. New Jersey Department of Transportation, Bureau of Materials.

4. U.S. EPA. 1992. Potential reuse of petroleum-contaminated soil: A directory of permitted recycling facilities. EPA/600/R-92/096. Washington, DC.

5. Blumenthal, M.H. 1993. Tires. In: Lund, H.F., ed. The McGraw-Hill recycling handbook. New York, NY: McGraw-Hill.

6. ARRA. 1992. 39 ARRA contractors save the taxpayer $664,184,838.50 in 1991 (press release). Asphalt Recycling and Reclaiming Association, 3 Church Circle, Suite 250, Annapolis, MD 21401.

7. Testa, S.M., and D.L. Patton. 1993. Soil remediation via environmentally processed asphalt. In: Hager, J.P., B.J. Hansen, J.F. Pusateri, W.P. Imrie, and V. Ramachandran, eds. Extraction and processing for the treatment and minimization of wastes. Warrendale, PA: The Minerals, Metals, and Materials Society. pp. 461-485.

8. U.S. EPA. 1990. Current status of the definition of solid waste as it pertains to secondary materials and recycling. Washington, DC. (November).

9. Leonard, L., A.J. Priest, J.L. Means, K.W. Nehring, and J.C. Heath. 1992. California hazardous waste minimization through alternative utilization of abrasive blast material. Proceedings of the 1992 HAZ-MAT West Annual Conference, Long Beach, CA (November).

10. Bouse, E.F., Jr., and J.W. Kamas. 1988. Update on waste as kiln fuel. Rock Products 91:43-47.

11. Bouse, E.F., Jr., and J.W. Kamas. 1988. Waste as kiln fuel, part II. Rock Products 91:59-64.

12. Heath, J.C., L. Karr, V. Novstrup, B. Nelson, S.K. Ong, P. Aggarwal, J. Means, S. Pomeroy, and S. Clark. 1991. Environmental effects of small arms ranges. NCEL Technical Note N-1836. Port Hueneme, CA: Naval Civil Engineering Laboratory.

13. Perry, R.H., and C.H. Chilton. 1984. Chemical engineers' handbook, 6th ed. New York, NY: McGraw-Hill.

14. Wills, B.A. 1985. Mineral processing technology, 3rd ed. New York, NY: Pergamon Press.

15. U.S. EPA. 1993. Technical resource document: Solidification/stabilization and its application to waste materials. EPA/530/R-93/012. Cincinnati, OH.

16. Queneau, P.B., and A.L. Troutman. 1993. Waste minimization charges up recycling of spent lead-acid batteries. HazMat World 6(8):34-37.

17. Timm, S.A., and K. Elliot. 1993. Secondary lead smelting doubles as recycling, site cleanup tool. HazMat World 6(4):64, 66.

18. U.S. EPA. 1992. The Superfund innovative technology evaluation program: Technology profiles, 5th ed. EPA/540/R-92/077. Washington, DC.

19. Paff, S.W., and B. Bosilovich. 1993. Remediation of lead-contaminated Superfund sites using secondary lead smelting, soil washing, and other technologies. In: Hager, J.P., B.J. Hansen, J.F. Pusateri, W.P. Imrie, and V. Ramachandran, eds. Extraction and processing for the treatment and minimization of wastes. Warrendale, PA: The Minerals, Metals, and Materials Society. pp. 181-200.

20. Horne, B., and Z.A. Jan. 1994. Hazardous waste recycling of MGP site by HT-6 high temperature thermal distillation. Proceedings of Superfund XIV

Conference and Exhibition, Washington, DC (November 30 to December 2, 1993). Rockville, MD: Hazardous Materials Control Resources Institute. pp. 438-444.

21. Miller, B.H. 1993. Thermal desorption experience in treating refinery wastes to BDAT standards. Incineration Conference Proceedings, Knoxville, TN (May).

22. U.S. EPA. 1992. Facility pollution prevention guide. EPA/600/R-92/088. Cincinnati, OH.

23. Krukowski, J. 1994. Thermal desorber treats oily sand for L.A. port authority. Poll. Eng. 26(4):71.

24. Villaume, J.F., P.C. Lowe, and D.F. Unites. 1983. Recovery of coal gasification wastes: An innovative approach. Proceedings of the National Water Well Association Third National Symposium on Aquifer Restoration and Ground-water Monitoring. Worthington, OH: Water Well Journal Publishing Company.

INDEX

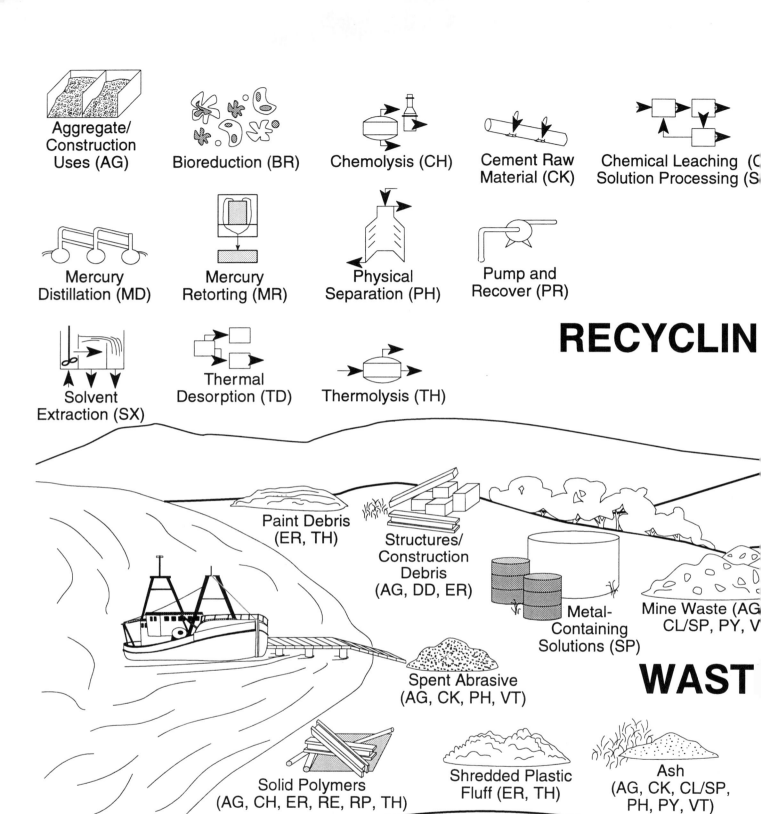

Aggregate/Construction Uses (AG)

Bioreduction (BR)

Chemolysis (CH)

Cement Raw Material (CK)

Chemical Leaching (C
Solution Processing (S

Mercury Distillation (MD)

Mercury Retorting (MR)

Physical Separation (PH)

Pump and Recover (PR)

Solvent Extraction (SX)

Thermal Desorption (TD)

Thermolysis (TH)

RECYCLIN

Paint Debris (ER, TH)

Structures/Construction Debris (AG, DD, ER)

Metal-Containing Solutions (SP)

Mine Waste (AG
CL/SP, PY, V

Spent Abrasive (AG, CK, PH, VT)

WAST

Solid Polymers (AG, CH, ER, RE, RP, TH)

Shredded Plastic Fluff (ER, TH)

Ash (AG, CK, CL/SP, PH, PY, VT)

Inorganics-Contaminated Sediments (CL/SP, PY, VT)

Inorganics-Contaminated Soils and Sludges (CL/SP, PY, VT)

Mercury Contamination (BR, CL/SP, MR, PH)

Organics-Contaminated Sediments (DE, ER, SX, TD)